Springer Theses

Recognizing Outstanding Ph.D. Research

Aims and Scope

The series "Springer Theses" brings together a selection of the very best Ph.D. theses from around the world and across the physical sciences. Nominated and endorsed by two recognized specialists, each published volume has been selected for its scientific excellence and the high impact of its contents for the pertinent field of research. For greater accessibility to non-specialists, the published versions include an extended introduction, as well as a foreword by the student's supervisor explaining the special relevance of the work for the field. As a whole, the series will provide a valuable resource both for newcomers to the research fields described, and for other scientists seeking detailed background information on special questions. Finally, it provides an accredited documentation of the valuable contributions made by today's younger generation of scientists.

Theses are accepted into the series by invited nomination only and must fulfill all of the following criteria

- They must be written in good English.
- The topic should fall within the confines of Chemistry, Physics, Earth Sciences, Engineering and related interdisciplinary fields such as Materials, Nanoscience, Chemical Engineering, Complex Systems and Biophysics.
- The work reported in the thesis must represent a significant scientific advance.
- If the thesis includes previously published material, permission to reproduce this must be gained from the respective copyright holder.
- They must have been examined and passed during the 12 months prior to nomination.
- Each thesis should include a foreword by the supervisor outlining the significance of its content.
- The theses should have a clearly defined structure including an introduction accessible to scientists not expert in that particular field.

More information about this series at http://www.springer.com/series/8790

Hanieh Fattahi

Third-Generation Femtosecond Technology

Doctoral Thesis accepted by
the Max Planck Institute for Quantum Optics, Germany

Springer

Author
Dr. Hanieh Fattahi
Attosecond Physics
Max Planck Institute for Quantum Optics
Garching
Germany

Supervisor
Prof. Ferenc Krausz
Max Planck Institute for Quantum Optics
Garching
Germany

ISSN 2190-5053 ISSN 2190-5061 (electronic)
Springer Theses
ISBN 978-3-319-20024-8 ISBN 978-3-319-20025-5 (eBook)
DOI 10.1007/978-3-319-20025-5

Library of Congress Control Number: 2015946753

Springer Cham Heidelberg New York Dordrecht London

Printed on acid-free paper

Springer International Publishing AG Switzerland is part of Springer Science+Business Media (www.springer.com)

Every man is the creature of the age in which he lives; very few are able to raise themselves above the ideas of the times.

Voltaire

The first principle is that you must not fool yourself, and you are the easiest person to fool.

Richard Feynman

To Asghar

Supervisor's Foreword

Ever since scientists have been able to control and record bursts of light shorter than the eye can see, we have been driven to invent new ways to take our capabilities to the next level, to allow us to peer deeper into the inner workings of all of the atoms and molecules that make up the world around us. A revolution in this line of research came with the invention of mode-locked dye lasers in the 1970s, which could produce pulses shorter than 1 ps, paving the way for the field of femtosecond science. This field underwent another transformation in the 1980s with the advent of broadband solid-state laser media such as titanium-doped sapphire, which boosted the peak and average powers of femtosecond sources by orders of magnitude, aided by the application of chirped pulse amplification to laser amplifiers to boost the achievable intensities to new limits, the ability to stabilize and control the carrier-envelope phase of modelocked lasers, and by the use of chirped multilayer mirrors to allow for the control of dispersion over wavelength ranges reaching an octave.

These were the key ingredients for us to reach into a new realm of light–matter interaction: attosecond physics, where the time resolution of our measurements has been able to reach the timescale of electronic motion. However, the more one works with a given technology, the more one becomes aware of its limitations: in the case of solid-state lasers, these are primarily the simultaneous availability of high peak and average powers, combined with ultra-broadband spectra, which would be necessary to, for example, form single-cycle bursts of extremely-intense light to drive the next generation of X-ray sources allowing for time-resolved investigation of the exact motion of electron density in complex molecules and solids.

So, how do we get there? To keep both high power and broad bandwidth presents a challenge, since increasing the emission bandwidth in a laser tends to lead to an increase in the quantum defect and subsequent efficiency loss. However, we can use nonlinear optics to overcome this limitation: optical parametric amplification (OPA) can yield both efficient energy transfer and broad gain bandwidths, while simultaneously making use of only virtual transitions in the gain medium, leaving very little heat behind in the crystal, paving the way for substantial increases in power. This is new territory for many of us in laser physics, though: how do we go from the understanding of the basic optical principles involved and go about making a practical, next-generation light source?

Now, perhaps, is a good time for me to introduce Hanieh Fattahi. She came to Germany from Iran in 2008 to pursue her interest in photonics and dreams of an academic career. After joining my group, she quickly gained a reputation among her colleagues as both as a thoroughly knowledgeable expert in nonlinear light propagation and optical system design, and as a patient and insightful teacher. These qualities are, I believe, well represented in her Ph.D. thesis, "Third-Generation Femtosecond Technology," which introduces the reader to the basic and not-so-basic details of building a high-energy, high-power ultrabroadband optical waveform synthesizer, such as the one we envision at our new laboratory, the Laboratory for Extreme Photonics (LEX), where she is taking on a leading role as a postdoctoral scientist.

Starting from the basics of the amplification of photons, she discusses all of the necessary ingredients for an OPA-based next-generation light source: the generation of a broadband seed, a reliable pump laser, and the amplifier itself. These are accompanied by numerous ideas and tips on how to improve efficiency to make good use of the high-energy pump pulses. Finally, the concluding chapter discusses the design of the synthesizer under construction at LEX, which provides a full picture of what we see as the frontier of ultrashort pulse generation to be explored in the coming years.

I trust that beginners to the field and experts alike will find something useful in the vast array of investigations that Dr. Fattahi performed during her time as a Ph.D. student and the detailed and accessible way in which they are described. It is an exciting time to be in the field of ultrafast and nonlinear optics, especially when one sees the excitement and intelligence of the next generation of researchers in the field.

Garching, Germany
July 2015

Prof. Ferenc Krausz
Director
Max-Planck-Institute of Quantum Optics
Chair for Experimental Physics, Laser Physics
Ludwig-Maximilians-Universität München

Abstract

Chirped pulse amplification in solid-state lasers is currently the method of choice for producing high-energy ultrashort pulses, having surpassed the performance of dye lasers over 20 years ago. The third generation of femtosecond technology based on short-pulse-pumped optical parametric chirped pulse amplification (OPCPA) holds promise for providing few-cycle pulses with terawatt-scale peak powers and kilowatt-scale-average powers simultaneously, heralding the next wave of atto-second and femtosecond science.

OPCPA laser systems pumped by near-1-ps pulses support broadband and efficient amplification of few-cycle pulses due to their unrivaled gain per unit length. This is rooted in the high threshold for dielectric breakdown of the nonlinear crystals for even shorter pump pulse durations. Concomitantly, short pump pulses simplify dispersion management and improve the temporal contrast of the amplified signal.

This thesis covers the main experimental and theoretical steps required to design and operate a high-power, high-energy, few-cycle OPCPA. This includes the generation of a broadband, high-contrast, carrier envelope phase (CEP)-stable seed, the practical use of a high-power thin-disk regenerative amplifier, its efficient use for pumping a multi-stage OPCPA chain and compression of the resulting pulses. A theoretical exploration of the concept and its extension to different modes of operation, including widely-tunable, high-power multi-cycle pulse trains, and ultrabroadband waveform synthesis is presented.

Finally, a conceptual design of a field synthesizer with multi-terawatt, multi-octave light transients is discussed, which holds promise for extending the photon energy attainable via high harmonic generation to several kilo-electronvolts, nourishing the hope for attosecond spectroscopy at hard-x-ray wavelengths.

Acknowledgments

This work was conducted in the Attosecond division of Max Planck Instittue of Quantum Optics in Garching, Germany. For sure I could not see the light at the end of the tunnel without the provided facility and environment in the chair of Prof. Krausz and Max Planck Institute and the help of my colleagues.

In this regard and in the very first place, I would like to thank my supervisor Prof. Ferenc Krausz for providing me a joyful and inspiring atmosphere to make this work to happen. I consider myself fortunate to have been part of such a wonderful research group and I especially truly enjoyed the freedom that I was given to pursue new ideas.

I would like to thank Dr. Nicholas Karpowicz, for his invaluable discussion in G2.22 (the best office in the world!), and his constant support and encouragement throughout my research.

I am also thankful to Dr. Thomas Metzger, Prof. Reinhard Kienberger, Dr. Alexander Apolonskiy, Dr. Shawn Sederberg, Dr. Wolfgang Schweinberger, Clemens Jakubeit, Dr. Martin Schultze, Dear Dr. Elisabeth Bothschafter, Dr. Alexander Schwarz, Dr. Volodymyr Pervak, Prof. Matthias Kling, Prof. Peter Hommelhoff, Dr. Vladislav Yakovlev, Roswitha Graf, Biddor Alsaif, Dr. Catherine Teisset, Dr. Oleg Pronin, Dr. Christoph Skrobol, Dr. Yunpei Deng, Dr. Xun Gu, Dr. Thomas Ganz, and Moritz Ueffing for their support and joyful scientific discussions and especially to Olga Razskazovskaya for preparing the article.

I owe my deepest dept of gratitude to my parents, Mohsen and Fatemeh, and to my siblings, Zohreh and Saeed for years of patience and encouragement.

And finally my sincerest gratefulness belongs to Asghar, the kindest, most supportive, and cleverest husband ever!

Contents

Acronyms

BOC	Balanced Optical Cross Correlator
CEP	Carrier Envelope Phase
COM	Center of Mass
CPA	Chirped Pulse Amplification
CW	Continues Wave
DFG	Difference Frequency Generation
DOPA	Degenerate Optical Parametric Amplification
DSLM	Dual Spatial Light Modulator
FWHM	Full Width at Half Maximum
GDD	Group Delay Dispersion
GVD	Group Velocity Dispersion
HCF	Hollow Core Fiber
HHG	High Harmonic Generation
KLM	Kerr Lens Mode-locked
LMA	Large Mode Area
MIR	Mid Infrared
NCPM	Noncritical Phase-Matching
NIR	Near Infrared
NOPA	Noncollinear Optical Parametric Amplification
OPA	Optical Parametric Amplification
OPCPA	Optical Parametric Chirped Pulse Amplification
PC	Pockels Cell
PCF	Photonic Crystal Fiber
PPLN	Periodically Poled $LiNbO_3$
PPSLT	Periodically Poled Stoichiometric $LiTaO_3$
SFG	Sum Frequency Generation
SHG	Second Harmonic Generation
SH-FROG	Second Harmonic—Frequency Resolved Optical Gating
SPM	Self-phase Modulation
SVEA	Slowly Varying Envelope Approximation
TAPIR	Temporal Allocation of Pump Intensity Remainder

TG-FROG	Transient Gating—Frequency Resolved Optical Gating
THG	Third Harmonic Generation
TPA	Two-Photon Absorption
UV	Ultraviolet
VIS	Visible
XFROG	Cross-Correlation Frequency Resolved Optical Gating
XPM	Cross-Phase Modulation
XPW	Crossed Polarized Wave Generation
XUV	Extreme Ultraviolet

Chapter 1
Introduction

It is a dream (one of mine, at least) to understand the hide and seek of photons, incognito messengers of information that let us explore the universe with quantum transitions. I am fascinated by the Greek and Persian philosophies of light, by the magic story of photons and what they make us perceive as reality. Physics and photonics are the grand attracting magnets as they make our further acquaintance with these tiny packets of energy possible. The story of the interconnection between life on earth and light spans a history of four billion years, when the first interaction between a bacterium and a photon took place. But the real success story of the human fascination with photons began a century ago.

Past The discovery of the principle of spontaneous and stimulated emission by Einstein (1917) is the root of almost all recent innovations in photonics. The suggested method for producing population inversion in the Ph.D. thesis of Fabrikant in 1939 and optical pumping by Kastler (1950), paved the route for demonstration of the first maser by Gordon et al. (1955). In 1958, Schawlow and Townes introduced the concept of the optical maser (laser) (Schawlow and Townes 1958) and 2 years later the first laser in a ruby crystal was demonstrated by Maiman (1960). The big question was on the table: What is this army of coherent photons oscillating at PHz frequencies with intensities around $10^{10}\,\mathrm{W/cm^2}$ good for?

One year later, lasers gave birth to the field of nonlinear optics by the generation of low-order optical harmonics in crystalline quartz (Franken et al. 1961). For electric field strengths less than the characteristic field strength of atoms, perturbative phenomena dominate. Here, the light perturbs the atoms, inducing a nonlinear (anharmonic) polarization and low-order harmonics may be generated.

In the 1970s, passively mode-locked dye lasers produced the first pulses shorter than 1 ps and gave birth to femtosecond technology (Ippen et al. 1972). Advances in laser oscillators led to pulse durations of a few tens of femtoseconds (Fork et al. 1981), while the poor energy storage capability of dye lasers limited the pulse energies to microjoules and the peak powers to megawatts (Knox et al. 1984). This first-generation femtosecond technology opened the door for direct time-domain investigations of processes that were once immeasurably fast, such as

H. Fattahi, *Third-Generation Femtosecond Technology*,
Springer Theses, DOI 10.1007/978-3-319-20025-5_1

molecular dynamics, chemical reactions, and phase transitions in condensed matter (Rowlinson 1969; Zewail 2000).

By the end of the 1980s, broadband solid-state lasers with large energy storage capabilities offered the potential for additional pulse shortening as well as boosting of the pulse energy and peak power by many orders of magnitude (Moulton 1986; Petrióevič et al. 1988).

Present Although the idea of optical stimulated emission was triggered by the same phenomena in the microwaves and masers, it took physicists fifteen years to exploit chirped-pulse amplification (Strickland and Mourou 1985), a well-known technique in microwave technology (Cook 1960), to solid-state lasers in particular Ti:sapphire-based systems (Nabekawa et al. 1998). At peak intensities on the order of 10^{14} W/cm^2 (Yanovsky et al. 2008; Sung et al. 2010), the laser field strength became comparable to the electric field experienced by a bound electron and nonlinear optics stepped into a new area, the so called non-perturbative regime.

In this regime electrons absorb photons in excess of the minimum required for ionization to occur (Agostini et al. 1979) and may lead to the generation of high-order harmonics, a coherent spectrum that could reach several kilo-electronvolts (Buth et al. 2013; Seres et al. 2005; Popmintchev et al. 2012). Ti:sapphire-based CPA systems combined with chirped multilayer mirrors form the second-generation of femtosecond technology, providing pulses with ultrahigh (petawatt) peak powers at relatively-low repetition rates (Chu et al. 2013) and moderate-peak-power (gigawatt) pulses at kilohertz repetition rates (Hädrich et al. 2013). Second-generation femtosecond technology paved the way for the emergence of entirely new research fields and technologies such as attosecond science (Krausz and Ivanov 2009) and has been the workhorse for the generation of isolated attosecond pulses in many laboratories around the world for more than a decade (Cavalieri et al. 2007).

Future Electron dynamics in atoms and ions take place on the time scale of attoseconds. The capabilities of the current high-harmonic generation (HHG) sources are limited to energies around a few hundred eV and to pulse durations of several tens of attoseconds. Attosecond pulses at high repetition rates, with photon energies as high as several kiloelectronvolts in the regime of hard X-rays, would enable the probing of electron dynamics in inner shells and 4D imaging of the electronic structure of matter with atomic temporal and spatial resolution, and open the door to time resolved intra-nuclear dynamics in addition to the ability to track the electron dynamics inside atoms.

Comprising optical parametric chirped-pulse amplification (OPCPA) (Dubietis et al. 1992) driven by terawatt-scale pulses from ytterbium-based lasers at kilowatt-scale average power, third-generation femtosecond technology will combine high (terawatt-scale) peak powers with high (kilowatt-scale) average powers in ultrashort optical pulse generation for the first time. This unprecedented parameter combination will allow the exploration of extreme nonlinearities in matter and extend ultrashort pulse generation from subnanometer to multimicrometer wavelengths at unforeseen flux levels, holding promise for yet another revolution in ultrafast science.

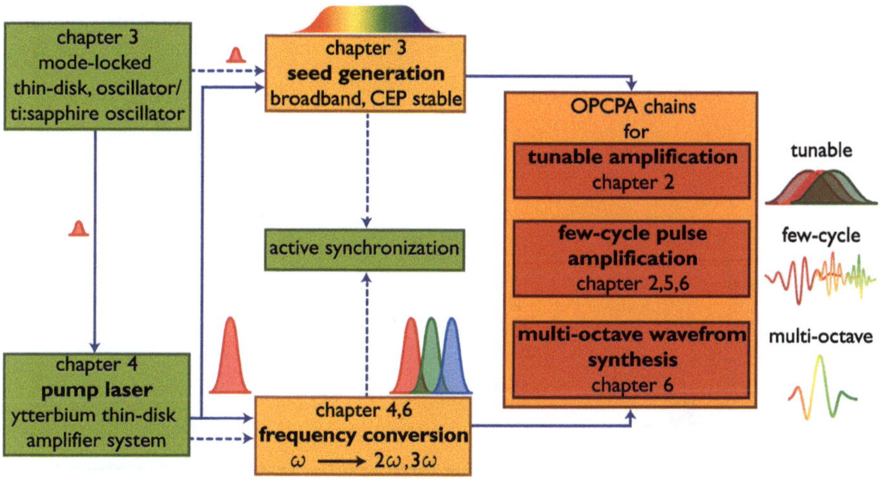

Fig. 1.1 Basic conceptual architecture of an OPCPA system. A sub-ps ytterbium laser oscillator or a Ti:sapphire oscillator, seeds the pump source. The broadband seed can either be generated from the output of the picosecond pump laser (*solid arrows*) or directly from the oscillator (*dashed arrows*). In the latter case an active temporal synchronization is needed between the pump and seed pulses of the OPCPA chain. This system can be operated to generate (i) widely tunable pulses of few-tens of femtoseconds duration, (ii) few-cycle pulses in different spectral ranges, or (iii) multi-octave controlled waveforms with sub-optical-cycle structure. These blocks are discussed in detail in the respective chapters as indicated in the figure (Fattahi et al. 2014a)

In this work it will be shown that diode-pumped ytterbium-doped thin-disk lasers offer a promising route to fulfill requirements to implement third-generation femtosecond technology in the conceptual architecture outlined in Fig. 1.1 and will outperform the first-and second-generations of femtosecond technology by reaching unprecedented peak and average powers, simultaneously (Fig. 1.2).

Besides a variety of operational modes, this next generation technology of waveform-controlled light offers an agile electromagnetic force whose strength can be increased from zero to maximum within hundred attoseconds and also kept nearly constant over several femtoseconds. The generation of synchronized UV and X-ray light at kHz repetition rates open the door for novel attosecond techniques such as UV transient absorption spectroscopy, XUV pump-probe spectroscopy, and attosecond X-ray diffraction. This high-energy, high-power field synthesizer will make it possible to understand multi-electron phenomena in solids and let us explore the ultimate limits of solid-state electronics. The new insights it leads to could bring about the next breakthroughs in optoelectronics needed to speed up information technology to light frequencies, reaching the world of light-wave electronics (Krausz and Stockman 2014).

This thesis combines the first experimental and theoretical steps towards the realization of a light source to fully exploit the potential of the third generation of femtosecond laser technology. It includes feasibility tests performed at the Laboratory for Attosecond Physics at the Max Planck Institute of Quantum Optics and

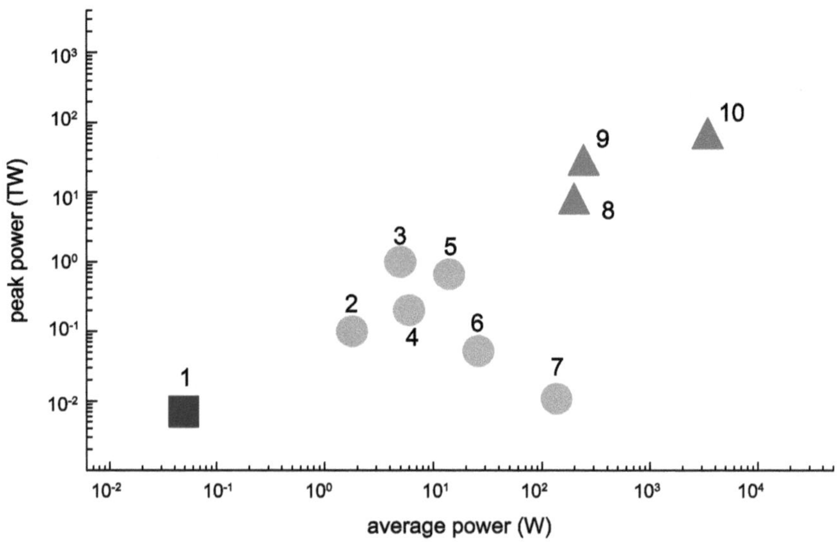

	τ_{pulse}	E_{pulse}	P_{peak}	$P_{average}$
system 8	5 fs	40 mJ	7.5 TW	200 W
system 9	1.7 fs	49 mJ	27 TW	245 W
system 10	5 fs	345 mJ	65 TW	3450 W

Fig. 1.2 Comparison between different generations of laser systems: The *blue dot* represent the performance of the first-generation of femtosecond technology in terms of the average and peak power (1, corresponds to (Rolland and Corkum 1986)), the green dots contain the second-generation of femtosecond technology (2–7, corresponds to (Nabekawa et al. 1998; Hädrich et al. 2013; Hentschel et al. 2000; Bohman et al. 2010; Chen et al. 2009; Matsushima et al. 2006)) and the *red dots* show the simulated results for OPCPA based on pump sources under development and envisioned. The table summarizes the predicted output parameters of these systems. The details of the simulation are presented in Appendix C (Fattahi et al. 2014a)

the Ludwig-Maximillians-University, and simple theoretical models to understand them. It also contains my personal interpretation of the phenomena which I dealt with in the laboratory, over the course of the last six years.

Outline of the thesis and summary of the main results Short-pulse pumped OPCPA systems are the most promising route to achieve high-contrast, few-cycle pulses scalable in peak and average power simultaneously.

It will be shown that the optimum duration of pump pulses in an OPCPA system to achieve a broad amplification gain and high conversion efficiency concurrently is of the order of 1 ps. This constitutes a trade-off between a high resistance to optical damage (decreasing for longer pulses) and a small temporal walk-off between pump and signal pulses relative to their duration (increasing for shorter pulses).

The realizability of this approach is subject to the availability of pump lasers delivering scalable high-energy near-1ps-pulses with near diffraction limited beam quality at repetition rates in the kHz to MHz range. Yb:YAG think disk lasers, containing all the above mentioned parameters are the most developed approach to achieve this goal and therefore formed the heart of all the experimentally demonstrated and designed OPCPA systems presented in this thesis.

Having the concept of short-pulse pumped OPCPA in place along with reliable pump sources, all the necessary ingredients of an OPCPA system will be discussed. This includes the full study of: alternative schemes for generation of multi-octave coherent seed sources for OPCPA, optimization of the performance of Yb:YAG thin-disk lasers for daily use, and presenting alternative schemes for increasing conversion efficiency and amplification bandwidth.

The gain bandwidth and therefore the attainable pulse duration in an OPCPA system is limited to the phase-matching range of the nonlinear crystal. To generate confined, sub-cycle pulses with controlled variability of the electric field, few-cycle pulses at different carrier frequencies can be synthesized temporally. The outcome of combining this concept with short-pulse-pumped OPCPA concept are sub-cycle pulses, scalable in peak and average power. It will be shown that few-cycle OPCPA pulses with mJ levels of energy in the visible, near-infrared, and mid-infrared ranges pumped by different harmonics of an Yb:YAG think-disk laser can be synthesized to form multi-TW sub-cycle waveforms. This extends the horizon of high-energy attosecond pulse generation to keV range.

Chapter 2
Summary Points

Chapter 2 gives an overview of optical parametric amplification and is dedicated to the theoretical description of the parametric amplification process. Numerical simulations for optimization of different parameters in OPCPA systems such as pump pulse duration, signal-to-pump energy ratio, and OPCPA phase are performed and discussed in detail. At the end of the chapter a general discussion about Yb:YAG thin-disk based OPCPA systems is presented. This section discusses the possibility of using fundamental, second- and third-harmonics of Yb:YAG lasers to pump few-cycle OPCPA systems in the visible, near-infrared, and mid-infrared. Here, different nonlinear crystals are analytically compared and the best candidates are suggested. The chapter is finalized by proposing a new scheme for a widely tunable OPCPA system (from ultraviolet to mid-infrared) based on Yb:YAG thin-disk lasers.

Main Results

In this chapter the behavior of OPCPA efficiency and amplification bandwidth versus the pump pulse duration (from fs to ns range) are theoretically investigated. It is shown the optimum pump pulse duration is in the range of 1–10 ps at FWHM. This study reveals that in this range, 1 ps pump pulses are favorable over 10 ps pulse duration, as they provide higher damage threshold on materials. It is theoretically demonstrated that increase in the OPCPA's pump intensity results in the raise of the small signal gain and its amplification bandwidth. This indicates another advantage of the short-pulse-pumped OPCPA concept.

Another important parameter in designing OPCPA system, is the input signal-to-pump energy ratio. The performed simulations in this chapter indicate, for low input seed energy, saturation of different frequency components happens at different length of the nonlinear medium. This results in backconversion of energy for some frequency components while the others have not reached saturation yet. The optimum signal-to-pump energy ratio of 10^{-3} is suggested for relatively uniform amplification while high conversion efficiency and amplification gain is achieved.

This chapter also includes simulations demonstrating that the phase-mismatch during the amplification process in an OPCPA system results in high-order phase. This additional phase is imprinted on the spectral phase of the signal. It is shown that increase in the amplification gain leads to increase in the OPA phase. This additional phase should be considered for Fourier limit pulse compression in OPCPA systems.

Finally a new scheme for an OPCPA-based tunable source is suggested. Here, a multi-octave seed pulse is temporally stretched to a pulse duration of several times the pump pulse duration. Using different harmonics of an Yb:YAG thin-disk laser, and by adjusting the temporal delay between seed and pump pulses, sub 30 fs pulses with mJ level of energy are generated. This table-top tunnable source is scalable in peak- and average-power with tunnablity in central frequency and temporal duration of pulses.

Chapter 3
Summary Points

Chapter 3 is dedicated to study alternative techniques for the generation of multi-octave broadband spectra suitable for seeding OPCPA systems. OPCPA seed pulses are required to have high temporal contrast, smooth spectral phase and preferably passively stable carrier envelope phase. This can be provided by broadband Ti:sapphire oscillators with nJ of energy or kerr-lens mode-locked Yb:YAG oscillators with μJ level of energy. Both require an additional active temporal synchronization to overcome temporal jitter between seed and pump pulses and they are limited in energy. As an alternative approach a broadband seed with several hundreds of μJ energy can be generated by employing a Ti:sapphire amplifier. However, this scheme also demands for an active temporal synchronization system. Generating seed pulses from a near-1 ps pulses of a Yb:YAG amplifier will eliminate the demand for any external synchronization system and provides the possibility of scaling the seed energy level to mJ.

Therefore in this chapter a scheme to generate an octave seed pulses from a 1-ps Yb:YAG oscillator by combining spectral broadening in large mode area (LMA) fiber and supercontinuum generation in bulk is studied. Later, supercontinuum generation of 30 fs pulses in a hollow-core-fibers (HCF) is discussed. The chapter concludes with a full experimental and theoretical study of cross-polarized-wave-generation (XPW) and difference frequency generation (DFG) serving as suitable techniques for OPCPA seed generation.

Main Results

In a cascaded broadening setup, 1 ps pulses centered at 1030 nm were spectrally broadened in an LMA fiber and temporally compressed to 30 fs and used for supercontinuum generation in YAG and YVO_4 crystal. In this approach the spectrum of the generated supercontinuum reached 500 nm. However the bottle neck of this approach is the LMA fiber which is limited to pump pulse energies of the order of several μJ. After that a full theoretical and experimental study on XPW generation as a technique for enhancing the temporal contrast of seed pulses is presented. It is shown that by employing a two-stage XPW, the efficiency of the process increase from 5 to 9 %. The demonstrated theoretical study and experimental evidence shows in an XPW signal which had reached energy saturation, the generated spectrum becomes modulated due to impact of the other third-order nonlinear processes along with the backconversion of energy from the XPW signal to the pump. These third-order nonlinear processes also result in spectral broadening and generation of new frequencies. It is experimentally shown, the input pump spectrum which spans from 600 to 1000 nm can be extended to 1400 nm in a single XPW stage using 1 mm-thick BaF_2. This makes XPW a useful technique for OPCPA seed generation.

A multi-octave spectrum spanning from visible to mid-infrared can be generated by spectral broadening of mid-infrared CEP-stable pulses in a gas or a gas-filled hollow core fiber. This method benefits from efficient generation of CEP-stable pulses in mid-infrared in a DFG process, as DFG pulses have high temporal contrast, smooth spectral phase and they are passively CEP-stable.

To increase the efficiency of a single-pulse DFG, a simple idea is implemented experimentally. Here the efficiency of a broadband DFG signal, pumped by a few-cycle near-infrared-to-visible pulses, is increased to 12 %. This increase in efficiency is achieved by employing a 42 μm-thick quartz wave plate before the 500 μm-thick BBO crystal used as the nonlinear medium. The wave plate optimizes the photon distribution on ordinary and extra-ordinary axes of the type I crystal. It will be shown that the type I DFG, has a higher efficiency due to its reduced sensitivity to the material dispersion, whereas in type II DFG, the generated signal is more stable.

This results will be used in Chap. 6 to suggest alternative seed-generation schemes for a multi-TW, super-octave field synthesizer.

Chapter 4
Summary Points

Chapter 4 presents the dynamics of chirped pulse amplification in Yb:YAG thin-disk regenerative amplifiers as the most promising pump source for OPCPA systems. Here numerous simulations are performed to investigate the accumulated nonlinear phase during the amplification process in a regenerative amplifier. This chapter also includes a full experimental and theoretical study about second harmonic generation of Yb:YAG thin-disk lasers.

Main Results

It will be shown experimentally and theoretically that by increasing the input seed energy of a thin-disk regenerative amplifier from 0.1 to 17 nJ a stable mode of operation without doubling and chaos can be achieved. Increasing the input seed energy also makes it possible to use less number of round trips in the amplifier and consequently reduce the accumulated nonlinear phase on the beam in the expense of using higher pump energy. Experimental and theoretical evidence indicate, that reproducible pumping of OPCPA systems requires a well-controlled balance between the demands of avoiding chaotic pulse train dynamics and providing a reproducible spectral phase.

The frequency doubling of an Yb:YAG thin-disk laser in type I BBO and LBO crystals is the second subject of this chapter. Up to 70 % of energy conversion is achieved in a 4 mm LBO and a 2 mm BBO at the pump peak intensity of 70 GW/cm^2. Although LBO has a lower nonlinear coefficient compared to BBO, it results in a similar conversion efficiency due to its smaller temporal walk-off. The achieved high conversion efficiency indicates the good quality of thin-disk amplifier in time and space.

Chapter 5
Summary Points

Chapter 5 is devoted to second harmonic-pumped, near-infrared (NIR) OPCPA systems. The chapter begins with a theoretical design study optimized for increasing OPCPA pump-to-signal conversion efficiency by recycling the residual pump energy in a cascaded OPCPA stage. Further optimization of efficiency is studied in a scheme consisting of controlled deposition of pump energy in different part of the spectrum. In the following, different methods for spectral bandwidth engineering by using different kinds of nonlinear crystal are experimentally and theoretically investigated.

Main Results

The conversion efficiency of a 3-stage few-cycle OPCPA system, centered at 800 nm is increased to an unprecedented value of 30 %. This is achieved by reusing the residual pump energy after the second OPCPA stage to pump the third OPCPA stage. The 2.5 mJ pulses with 7.5 W of average power are temporally compressed to 10 fs employing a chirped-mirror compressor providing 0.23 TW of peak power.

In the second approach the possibility of increasing the OPCPA conversion efficiency by controlling the allocation of pump energy on different spectral components is theoretically discussed. In this technique seed pulses are temporally stretched to longer than pump pulse duration. The pump is reused after each OPCPA stage and the pump energy in each OPCPA stage is deposited on different part of the seed spectrum by adjusting the relative temporal delay between pump and signal. This technique also leads to an smooth amplified spectrum, which is demonstrated experimentally.

And finally, near-1-cycle pulses with 1.8 mJ of energy are generated in a 3-stage noncollinear OPCPA system centered at 1 μm. Here by using a 2 mm LBO, 2 mm BBO, and 3 mm LBO crystal, the amplification bandwidth is extended to support near-1-cycle pulses. Using a chirped-mirror compressor resulted in 9 fs pulses.

And finally in an all LBO 3-stage OPCPA system, generated spectrum from the XPW setup shown in Chap. 3 is amplified to 1.2 mJ of energy. Here the amplified spectrum is smooth without any sharp features, indicating the enhanced contrast of the amplified pulses.

Chapter 6
Summary Points

Chapter 6 consists of a design study of a multi-channel OPCPA system based on Yb:YAG thin-disk lasers. Here, different harmonics of the laser are used to pump different few-cycle OPCPA channels in visible, near-infrared, and mid-infrared. Next, the generation of a variety of ultra-short wave transients from the three-channel OPCPA-based synthesizer and their optimization as a driving field for HHG in order to reach keV photon energies is discussed theoretically.

Main Results

The full design study of a multi-TW, multi-octave OPCPA based field synthesizer is shown in this chapter. An Yb:YAG thin-disk laser with 200 mJ of energy operating at 5 kHz is considered as a pump source for this system. In a cascaded second-order nonlinear stage, 40 mJ pulses at 343 nm, 74 mJ pulses at 515 nm are generated, leaving 86 mJ energy at 1030 nm.

These pulses are used to pump three OPCPA chains employing BBO, LBO, and $LiNbO_3$, alternatively. The amplified spectrum in 343 nm pumped OPCPA chain spans from 450 to 650 nm and contains 7.1 mJ of energy. In the near-infrared channel the 680–1300 nm amplified spectrum contains 22.7 mJ of energy and the spectrum in the mid-infrared channel from 1700 to 2800 nm has 19.2 mJ of energy.

By synthesizing these amplified spectra after their temporal compression, and varying temporal delays between the three different channels, a variety of confined sub-cycle multi-TW waveforms could be generated.

In a series of simulations, the advantages of controlled waveforms over one-cycle and multi-cycle pulses for high harmonic generation in terms of generated photon cut-off energy is shown. It is shown by optimizing the energy ratio between the visible, near-infrared, and mid-infrared channels in favor of mid-infrared channel, the photon cut off energy can be extended dramatically in the high harmonic generation process. Simulation results indicate, chirping the waveform by shifting the blue channel to the front of the pulse also leads to an increase of the photon cut off energy. This could be understood by the role of blue photons in the leading edge of the pulse in ionization and the longer wavelength photons in the trailing edge of the waveform in electron recombination and high harmonic generation.

Comparison between the ionization probability of a 4 fs pulse at 800 nm and a synthesized waveform at 1 μm, shows lower ionization rate for the waveform for the case of similar cut-off energy. This demonstrate another advantage of a waveform over few-cycle pulses for high harmonic generation process.

Chapter 7
Summary Points

Finally, Chap. 7 is followed by a summary of the findings and perspectives for the future.

Appendix A and D
Summary Points

Appendix A and D, cover two important subjects. Spectral interferometry as an important tool for characterizing chirped-mirror compressor and supercontinuum generation in bulk material pumped by 1 ps pulses.

References

Einstein A (1917) Zur Quantentheorie der Strahlung. Physikalische Zeitschrift 18:121–128

Kastler A (1950) Quelques suggestions concernant la production optique et la détection optique d'une inégalité de population des niveaux de quantifigation spatiale des atomes. Application à l'expérience de Stern et Gerlach et à la résonance magnétique. J Phys Rad 11:255–265

Gordon J, Zeiger H, Townes C (1955) The maser-new type of microwave amplifier, frequency standard, and spectrometer. Phys Rev 99(4):1264–1274

Schawlow A, Townes C (1958) Infrared and optical masers. Phys Rev 112(6):1940–1949

Maiman TH (1960) Stimulated optical radiation in ruby. Nature 187(4736):493–494

Franken P, Hill A, Peters C, Weinreich G (1961) Generation of optical harmonics. Phys Rev Lett 7(4):118–119

Ippen E, Shank CV, Dienes A (1972) Passive mode locking of the cw dye laser. Appl Phys Lett 21(8):348–350

Fork RL, Greene BI, Shank CV (1981) Generation of optical pulses shorter than 0.1 psec by colliding pulse mode locking. Appl Phys Lett 38(9):671–672

Knox WH, Downer MC, Fork RL, Shank CV (1984) Amplified femtosecond optical pulses and continuum generation at 5-kHz repetition rate. Opt Lett 9(12):552–554

Rowlinson JS (1969) Thomas Andrews and the critical point. Nature 224(5219):541–543

Zewail AH (2000) Femtochemistry: atomic-scale dynamics of the chemical bond †. J Phys Chem A 104(24):5660–5694

Moulton PF (1986) Spectroscopic and laser characteristics of Ti:Al_2O_3. J Opt Soc Am B 3(1):125–133

Petri6evič V, Gayen SK, Alfano RR, Yamagishi K, Anzai H, Yamaguchi Y (1988) Laser action in chromium-doped forsterite. Appl Phys Lett 52(13):1040–1042

Strickland D, Mourou G (1985) Compression of amplified chirped optical pulses. Opt Commun 55(6):447–449

Cook C (1960) Pulse compression-key to more efficient radar transmission. Proc IRE 48(3):310–316

Nabekawa Y, Kuramoto Y, Togashi T, Sekikawa T, Watanabe S (1998) Generation of 0.66-TW pulses at 1 kHz by a Ti:sapphire laser. Opt Lett 23(17):1384–1386

Yanovsky V, Chvykov V, Kalinchenko G, Rousseau P, Planchon T, Matsuoka T, Maksimchuk A, Nees J, Cheriaux G, Mourou G, Krushelnick K (2008) Ultra-high intensity-300-TW laser at 0.1 Hz repetition rate. Opt Exp 16(3):2109–2114

Sung J, Lee S, Yu T, Jeong T, Lee J (2010) 0.1 Hz 1.0 PW Ti: sapphire laser. Opt Lett 35(18):3021–3023

Agostini P, Fabre F, Mainfray G, Petite G, Rahman N (1979) Free-free transitions following six-photon ionization of xenon atoms. Phys Rev Lett 42(17):1127–1130

Buth C, He F, Ullrich J, Keitel CH, Hatsagortsyan KZ (2013) Attosecond pulses at kiloelectronvolt photon energies from high-orderharmonic generation with core electrons. Phys Rev A 88(3):033848–033857

Seres J, Seres E, Verhoef AJ, Tempea G, Streli C, Wobrauschek P, Yakovlev V, Scrinzi A, Spielmann C, Krausz F (2005) Laser technology: source of coherent kiloelectronvolt X-rays. Nature 433(7026):596

Popmintchev T, Chen M-C, Popmintchev D, Arpin P, Brown S, Alisauskas S, Andriukaitis G, Balciunas T, Mücke OD, Pugzlys A, Baltuska A, Shim B, Schrauth SE, Gaeta A, Hernández-García C, Plaja L, Becker A, Jaron-Becker A, Murnane MM, Kapteyn HC (2012) Bright coherent ultrahigh harmonics in the keV X-ray regime from midinfrared femtosecond lasers. Science (New York, N.Y.) 336(6086):1287–1291

Chu Y, Liang X, Yu L, Xu Y, Xu L, Ma L, Lu X, Liu Y, Leng Y, Li R, Xu Z (2013) High-contrast 2.0 Petawatt Ti:sapphire laser system. Opt Exp 21(24):29231–29239

Hädrich S, Klenke A, Hoffmann A, Eidam T, Gottschall T, Rothhardt J, Limpert J, Tünnermann A (2013) Nonlinear compression to sub- 30-fs, 0.5 mJ pulses at 135 W of average power. Opt Lett 38(19):3866–3869

Dudley JM, Coen S (2006) Supercontinuum generation in photonic crystal fiber. Rev Modern Phys 78(4):1135–1184

Cavalieri AL, Goulielmakis E, Horvath B, Helml W, Schultze M, Fieß M, Pervak V, Veisz L, Yakovlev VS, Uiberacker M, Apolonski A, Krausz F, Kienberger R (2007) Intense 1.5-cycle near infrared laser waveforms and their use for the generation of ultra-broadband soft-X-ray harmonic continua. New J Phys 9(7):2421–2429

Dubietis A, Jonušauskas G, Piskarskas A (1992) Powerful femtosecond pulse generation by chirped and stretched pulse parametric amplification in BBO crystal. Opt Commun 88(4–6):437–440

Krausz F, Ivanov M (2009) Attosecond physics. Rev Mod Phys 81(1):163–234

Krausz F, Stockman MI (2014) Attosecond metrology: from electron capture to future signal processing. Nat Photon 8(3):205–213

Fattahi H, Barros HG, Gorjan M, Nubbemeyer T, Alsaif B, Teisset CY, Schultze M, Prinz S, Haefner M, Ueffing M, Alismail A, Vámos L, Schwarz A, Pronin O, Brons J, Geng XT, Arisholm G, Ciappina M, Yakovlev VS, Kim D-E, Azzeer AM, Karpowicz N, Sutter D, Major Z, Metzger T, Krausz F (2014a) Third-generation femtosecond technology. Optica 1(1):45–63

Rolland C, Corkum P (1986) Amplification of 70 fs pulses in a high repetition rate XeCl pumped dye laser amplifier. Opt Commun 59(1):64–68

Hentschel M, Cheng Z, Krausz F, Spielmann C (2000) Generation of 0.1-TW optical pulses with a single-stage Ti:sapphire amplifier at a 1-kHz repetition rate. Appl Phys B 70(S1):S161–S164

Bohman S, Suda A, Kanai T, Yamaguchi S, Midorikawa K (2010) Generation of 5.0 fs, 5.0 mJ pulses at 1kHz using hollow-fiber pulse compression. Opt Lett 35(11):1887–1889

Chen S, Chini M, Wang H, Yun C, Mashiko H, Wu Y, Chang Z (2009) Carrier-envelope phase stabilization and control of 1 kHz, 6 mJ, 30 fs laser pulses from a Ti:sapphire regenerative amplifier. Appl Opt 48(30):5692–5695

Matsushima I, Yashiro H, Tomie T (2006) 10 kHz 40 W Ti:sapphire regenerative ring amplifier. Opt Lett 1(13):2066–2068

Chapter 2
How to Amplify Photons

Stimulated emission and optical parametric amplification (OPA) lead to efficient amplification of photons. In stimulated emission, the amplification is limited by amplified spontaneous emission (ASE) in the gain medium, and the amplification bandwidth to the energy level structure of a laser active medium and gain narrowing. In this process the energy is stored in the medium.

OPA however is an instantaneous process with a high single pass gain. Here, the energy of a typically narrowband, high-energy pulse is transferred to a possibly broadband, low-energy pulse in a nonlinear crystal. In the absence of low-energy photons and at high intensities, high-energy photons decay spontaneously to two photons with lower energy via the emission of two photons. The process is called parametric fluorescence and its amplification in subsequent OPA media limits the achievable gain.

OPA provides broad amplification bandwidth which can be tuned by crystal type, crystal thickness, temperature, and the geometry of the interacting beams. To increase the conversion efficiency, the low-energy seed pulses should be temporally stretched to the time window of high-energy pump pulses. This is therefore called optical parametric chirped pulse amplification (OPCPA), combining OPA with chirped pulse amplification (CPA), but with different objective in the stretching of the seed pulses than in standard laser CPAs.

As here, since no energy is stored in the material, the thermal load on the crystal is significantly smaller. However, at several hundreds of kHz repetition rate, the small absorption of idler or pump can become significant and active cooling of the nonlinear medium is needed.

One of the most important properties of OPCPA is its simultaneous scalability in peak- and average-power. This unique property demands for powerful, high-energy, high-power pump sources with optimum pulse duration. For long-pulse-pumped OPCPAs, the intensity damage threshold of materials is on the order of $30\,\mathrm{GW/cm^2}$. Therefore for sufficient gain, thick crystals have to be used, limiting the amplification bandwidth. Decreasing the pump pulse duration results in the increase of the intensity damage threshold. However for pump pulses below $\sim$500 fs the OPCPA

H. Fattahi, *Third-Generation Femtosecond Technology*,
Springer Theses, DOI 10.1007/978-3-319-20025-5_2

conversion efficiency drops due to the increase in group velocity mismatch between the interacting pulses.

This chapter is devoted to the fundamentals of the OPCPA process. In the first section numerical simulations are performed to find the optimum pump pulse duration, optimum signal-to-pump energy ratio, and to study the induced spectral phase in an OPCPA system. In Sect. 2.2 a general concept for an OPCPA system pumped by Yb:YAG thin-disk amplifiers is reviewed and at the end a scheme for a widely tunable source based on OPCPA is presented.

2.1 Optical Parametric Amplification (OPA)

When an electric field interacts with a material, a nonlinear polarization is induced. This polarization depends on the strength of the electric field (E) and the polarization response of the material, called the electric susceptibility, χ.

Using a Taylor series expansion of the polarization in terms of the electric field and assuming an instantaneous response, the induced polarization is typically defined as:

$$P = \epsilon_0 (E + \chi^{(1)} E + \chi^{(2)} E^2 + \chi^{(3)} E^3 + \cdots). \tag{2.1}$$

OPA is a nonlinear process in the perturbative regime which is described by the second-order susceptibility ($\chi^{(2)}$). This non-resonant process can be simply explained by considering the photon-picture of light. In this process, the medium is excited to a virtual level with a higher energy by a photon with the energy of $\hbar\omega_p$, where ω_p is the angular frequency of the photon. Coupling to a virtual level represents a transient, polarized state of the system. After a non-resonant field interacts with the medium, the population returns to the ground state nearly instantaneously, with a slight delay determined by how far the virtual level is from a real level. In a medium with a nonzero $\chi^{(2)}$, a photon with lower energy of $\hbar\omega_s$ (ω_s represents the angular frequency of the low-energy photon) may stimulate the decay, and consequently two photons, with energies of $\hbar\omega_s$ and $\hbar\omega_p - \hbar\omega_s$ will be generated (Fig. 2.1).

Considering the wave picture of light, when two electric fields with the carrier frequencies of ω_p and ω_s interact, the envelope of the intensity resulting from their superposition has a modulation with the frequency of $\omega_p - \omega_s$. In the interaction of this field with the medium, a new electric field at the carrier frequency of $\omega_i = \omega_p - \omega_s$ is generated. The emission of these photons is caused by the presence of a small perturbation on the envelope of the input field, and results in the growth of this feature via energy transfer from the higher frequency field. In a *gedanken* experiment, the spectrum of the spontaneous emission in a very thin medium could span from ω_p to zero. However, in reality the bandwidth of the spectrum is limited by absorption and the phase-matching conditions of the material. The concept of phase-matching is based on two unbreakable laws of physics: the conservation of energy and momentum. The conservation of energy, for three interacting beams with

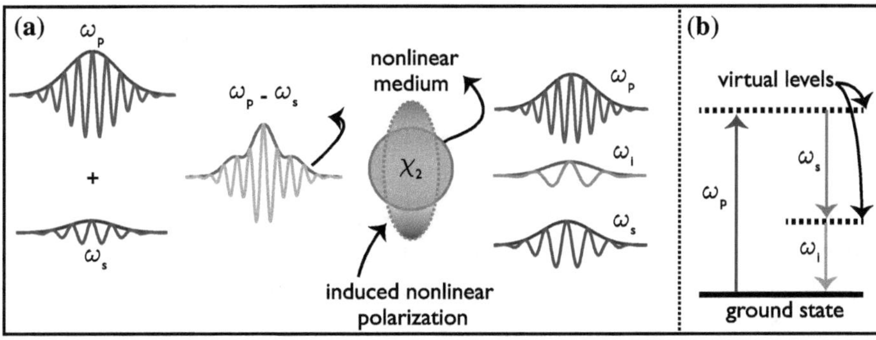

Fig. 2.1 **a** The wave-picture and **b** the photon-picture of OPA process

the central frequency of ω_p, ω_s, and ω_i (pump, signal, and idler respectively), is given by:

$$\omega_p - \omega_s - \omega_i = 0. \tag{2.2}$$

In reality, a pulse with a single frequency does not exist. Allowing a certain spectral bandwidth for each of these frequencies, the condition becomes:

$$\omega_p \pm \Delta\omega_p = \omega_s \pm \Delta\omega_s + \omega_i \pm \Delta\omega_i. \tag{2.3}$$

From the conservation of momentum:

$$\vec{k_p} - \vec{k_s} - \vec{k_i} = 0, \tag{2.4}$$

and including the spectral bandwidth:

$$\vec{k_p} \pm \Delta\vec{k}_p - (\vec{k_s} \pm \Delta\vec{k}_s) - (\vec{k_i} \pm \Delta\vec{k}_i) = \Delta\vec{k}, \tag{2.5}$$

from Eqs. (2.4) and (2.5):

$$\Delta\vec{k_p} - \Delta\vec{k_s} - \Delta\vec{k_i} = \Delta\vec{k}. \tag{2.6}$$

By substituting Eq. (2.3) in Eq. (2.6) and the Taylor expansion to first order:

$$\frac{\partial k_p}{\partial \omega_p}\Delta\omega_p - \frac{\partial k_s}{\partial \omega_s}\Delta\omega_s - \frac{\partial k_i}{\partial \omega_i}\Delta\omega_i = \Delta k, \tag{2.7}$$

$$\frac{\partial k_p}{\partial \omega_p}\Delta\omega_p - \frac{\partial k_s}{\partial \omega_s}\Delta\omega_s - \frac{\partial k_i}{\partial \omega_i}\Delta\omega_p + \frac{\partial k_i}{\partial \omega_i}\Delta\omega_s = \Delta k, \tag{2.8}$$

$$\Delta\omega_p(\frac{\partial k_p}{\partial\omega_p} - \frac{\partial k_i}{\partial\omega_i}) + \Delta\omega_s(\frac{\partial k_i}{\partial\omega_i} - \frac{\partial k_s}{\partial\omega_s}) = \Delta k, \quad (2.9)$$

$$\Delta\omega_p(\frac{1}{v_{gp}} - \frac{1}{v_{gi}}) + \Delta\omega_s(\frac{1}{v_{gi}} - \frac{1}{v_{gs}}) = \Delta k. \quad (2.10)$$

where v_{gj} for $j = p, s, i$ is the group velocity of the interacting waves and is defined by $\frac{d\omega_j}{dk_j}$. Equation (2.10) suggests that the phase-mismatch in the OPA process originates from the different group velocities of the interacting waves. It also suggests for the case of a very narrow-band pump beam, where $\Delta\omega_p \ll \Delta\omega_s$, the phase-mismatch is dominated by the group velocity difference between the signal and idler waves.

For the case of a broadband pump, Eq. (2.10) is zero when $v_{gi} = v_{gs} = v_{gp}$ or when:

$$\Delta\omega_p = \Delta\omega_s \left[\frac{\frac{1}{v_{gs}} - \frac{1}{v_{gi}}}{\frac{1}{v_{gp}} - \frac{1}{v_{gi}}}\right]. \quad (2.11)$$

Equation (2.11) indicates that the phase-matching bandwidth can be increased by using pump pulses with broader spectral bandwidth. This can be understood intuitively since in a broadband pump more pairs of frequencies meet the condition for zero phase-mismatch in comparison to a narrowband or a CW pump.

Under the assumption of an undepleted pump, the parametric gain bandwidth at full width at half maximum (FWHM) is:

$$\Delta\nu = \frac{2\,(\ln 2)^{\frac{1}{2}}}{\pi}\left(\frac{\Gamma}{L}\right)^{\frac{1}{2}}\frac{1}{|\delta_{si}|}, \quad (2.12)$$

where δ_{si} is the group velocity mismatch between signal and idler, and Γ is:

$$\Gamma^2 = \frac{\omega_i\omega_s d_{eff}^2 \mid A_p \mid^2}{n_i n_s c^2} = \frac{2d_{eff}^2\omega_i\omega_s}{c^3\epsilon_0 n_i n_s n_p} I_p = \frac{8\pi^2 d_{eff}^2 I_p}{n_i n_s n_p \lambda_i \lambda_s \epsilon_0 c}, \quad (2.13)$$

where, L is the length of the nonlinear medium, A_p is the envelope of the pump pulse, I_p is the pump intensity, d_{eff} is the effective nonlinear optical coefficient, and n_j and λ_j, $j = i, p, s$ indicate the refractive index of the nonlinear medium and the wavelength of the idler, signal, and pump, respectively.

Equation (2.12) shows that the gain bandwidth is inversely proportional to the group velocity mismatch between the signal and idler. In the special case where the central frequency of the signal is exactly half that of the pump, the generated idler covers the same spectral range as the seed. In this case, $v_{gi} = v_{gs}$ and Eq. (2.12) loses validity and Eq. (2.6) must be extended to second order (Cerullo and De Silvestri 2003):

$$\Delta\nu = \frac{2\,(\ln 2)^{\frac{1}{4}}}{\pi}\left(\frac{\Gamma}{L}\right)^{\frac{1}{4}}\frac{1}{\left|\frac{\partial^2 k_s}{\partial\omega_s^2}+\frac{\partial^2 k_i}{\partial\omega_i^2}\right|}. \tag{2.14}$$

In this case the gain bandwidth is proportional to the group velocity dispersion of the idler and signal. The Kramers-Kronig relation indicates that the frequency-dependent refractive index of materials changes dramatically between the electronic absorption band and phonon absorption bands of materials but at a saddle point between these two absorption regions, the group velocity dispersion (GVD) of materials becomes zero. For the frequencies in the vicinity of the saddle point, the deviation in dispersion, and therefore their temporal walk-off, is at its minimum.

This property of materials is utilized to phase-match and amplify a broad region of frequencies Eq. (2.14), in the so-called degenerate OPA (DOPA). In addition to parametric amplification in DOPA, second harmonic generation of the signal is also phase-matched, which causes an unwanted backward flow of energy from the signal to the pump. Ultimately, this modifies the shape of the amplified spectrum, limits the conversion efficiency, and can distort the spectral phase.

For frequencies away from the degeneracy point, the temporal walk-off depends on the group velocity Eq. (2.12) and can be compensated by exploiting a noncollinear geometry (Gale et al. 1995, 1998). Here, phase-matching is achieved by adding a noncollinear angle (α) between pump and signal, providing a longer temporal overlap between the pulses. In this case the group velocity of the signal is equal to the projection of the group velocity of the idler along the signal direction. In comparison with collinear geometry where signal and idler become quickly separated, in noncollinear geometry all three beams stay effectively overlapped. The idler generated in this process becomes angularly dispersed to fulfill the conservation of momentum Fig. 2.2 (Riedle et al. 2000). Then the phase-matching condition becomes:

$$v_{gs} = v_{gi}\cos\Omega, \tag{2.15}$$

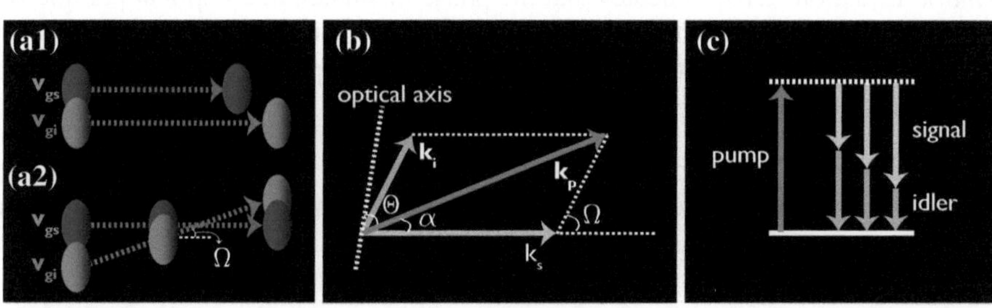

Fig. 2.2 **a1** The group velocity of the signal and idler pulses in a collinear OPA. Due to dispersion, pulses with longer wavelengths propagate faster than pulses with shorter wavelengths. **a2** This group velocity mismatch is compensated in a noncollinear geometry. **b** The phase-matching geometry for a noncollinear OPA. **c** The energy diagram for a broadband optical parametric amplifier

where Ω is the angle between the signal and idler wave vectors. Due to the angle between the interacting beams, this technique is called noncollinear optical parametric amplification (NOPA).

In DOPA and NOPA, ordinary and extraordinary axes of birefringence crystals are used. Here, the phase matching angle (θ) is defined as the angle between the optical axis of the crystal and the wave vector of the propagating beam. Two types of phase-matching can be realized in birefringent crystals. In type I phase-matching the two input pulses with lower frequencies have the same polarization, whereas in type II, they have different polarizations. A full study about this subject can be found in (Dmitriev et al. 1999).

Besides the mentioned phase-matching techniques, quasi phase-matching can be used to engineer and increase the coherence length in a nonlinear process. Here the phase-mismatch is compensated by inverting the orientation of the crystal axis, with a period equal to twice the coherence length. This reverses the sign of d_{eff} as soon as the field amplitude of the newly generated wave starts to decrease. The broadest spectral bandwidth can be achieved for frequencies around the degeneracy point of the material by poling a crystal with the period which compensates the wave vector mismatch of a frequency ω detuned from the zero group-delay dispersion (GDD) frequency.

Unlike DOPA and NOPA where ordinary and extra-ordinary axis of a birefringent crystal is exploited to compensate for phase-mismatch, quasi phase-matching permits the use of the same polarization for all three interacting beams. This often corresponds to the strongest element of the nonlinear tensor of the material which increases the conversion efficiency of the process. In the fabrication of periodically poled crystals, a permanent electrical polarization is induced into a ferroelectric material by applying an electric field. Therefore the quasi phase-matched crystals are limited in aperture size, as it is not yet possible to apply a uniform field on a crystal bigger than a certain size, generally a few millimeters. Consequently, this type of phase-matching is not applicable for high-energy OPCPA systems.

Remark Equation (2.10) indicates that the phase mismatch in an optical parametric amplifier is due to the group velocity difference between the interacting beams. It is also shown theoretically that using pulses with broad spectral bandwidth to pump OPCPAs result in a broader gain bandwidth.

To overcome the group velocity mismatch during the amplification, and to achieve broadband amplification, several different methods such as: noncollinear OPA, degenerate OPA, quasi-phase-matched OPA, and OPA in the Fourier domain (Schmidt et al. 2014), can be employed.

Intensity and Gain

Beginning with Maxwell's curl equation, the nonlinear wave equation assuming the instantaneous response for polarization can be obtained (Boyd 2003):

$$\nabla^2 E - \frac{\epsilon}{c^2}\frac{\partial^2 E}{\partial t^2} = \frac{1}{\epsilon_0 c^2}\frac{\partial^2 P^{nl}}{\partial t^2}. \tag{2.16}$$

For the phenomena studied in this thesis, it is assumed that the propagation length of interacting electromagnetic waves is much longer than their wavelength. The waves can be treated as light rays or plane waves, constant frequency waves with infinite parallel planes of constant peak-to-peak amplitude normal to the phase velocity vector. Sine and cosine functions meet the above criteria. For a linearly polarized wave propagating in the z direction, one can write:

$$E(z,t) = \Re[A(z)\exp(i\omega t - ikz)], \tag{2.17}$$

and the nonlinear polarization:

$$P^{nl}(z,t) = \Re[P^{nl}(z)\exp(i\omega t - ik_p z)]. \tag{2.18}$$

By substituting Eqs. (2.17) and (2.18) in Eq. (2.16):

$$\left[\frac{\mathrm{d}^2 A(z)}{\mathrm{d}z^2} - 2ik\frac{\mathrm{d}A(z)}{\mathrm{d}z} - A(z)k^2 + \frac{\epsilon\omega^2}{c^2}A(z)\right]\exp(i\omega t - ikz)$$
$$= -\frac{\omega^2}{\epsilon_0 c^2}P^{nl}(z)\exp(i\omega t - ik_p z). \tag{2.19}$$

As the envelope of the wave varies slowly in time and space in comparison to its wavelength, the slowly varying envelope approximation (SVEA) is often employed:

$$\frac{\mathrm{d}^2 A}{\mathrm{d}z^2} \ll 2k\frac{\mathrm{d}A}{\mathrm{d}z}. \tag{2.20}$$

Taking into account that:

$$k = \frac{n\omega}{c}, \tag{2.21}$$

Equation (2.19) becomes:

$$\frac{\mathrm{d}A}{\mathrm{d}z} = -i\frac{\mu_0 c\omega}{2n}P^{nl}\exp(ikz - ik_p z), \tag{2.22}$$

where n is the refractive index at frequency ω. This general equation shows that the amplitude variation of the propagating wave is caused by the nonlinear source term on the right hand side of Eq. (2.22). As OPA is a second-order nonlinear process, Eq. (2.1) can be written as:

$$P_i^{nl} = \epsilon_0 \chi_{ijk}^{(2)} E_j E_k, \tag{2.23}$$

where *i*,*j*, and *k* refers to three interacting waves, at frequencies ω_p, ω_s, and ω_i and $\chi^{(2)}$ is the second-order susceptibility tensor. By assuming the simple case of collinear geometry the following coupled wave equations can be derived from Eqs. (2.23) and (2.22):

$$\frac{dA_i}{dz} = -i\frac{\omega_i d_{eff}}{n_i c} A_s^* A_p \exp(-i\Delta k z), \tag{2.24}$$

$$\frac{dA_s}{dz} = -i\frac{\omega_s d_{eff}}{n_s c} A_i^* A_p \exp(-i\Delta k z), \tag{2.25}$$

$$\frac{dA_p}{dz} = -i\frac{\omega_p d_{eff}}{n_p c} A_i^* A_s \exp(i\Delta k z), \tag{2.26}$$

where d_{eff} is the effective nonlinear optical coefficient, which depends on the propagation direction and polarization of the three beams, and $\Delta k = k_p - k_s - k_i$ is the phase-mismatch.

Assuming an initial signal intensity A_{s0}, zero initial idler intensity $A_{i0} = 0$ and neglecting pump depletion $A_p \cong const.$, the coupled wave equation for a nonlinear crystal of length L can be solved for the intensity gain G:

$$G = \frac{I_s}{I_{s0}} = 1 + [\frac{\Gamma}{g}\sinh(gL)]^2, \tag{2.27}$$

where g is the small signal gain:

$$g = \sqrt{(\Gamma^2 - (\frac{\Delta k}{2})^2)}. \tag{2.28}$$

It can be seen from Eq. (2.28) that the small signal gain depends on the phase-mismatch between the interacting beams and Γ, where Γ is proportional to the pump intensity I_p. This suggests that there are two parameters which have the main role in the shape of the amplified spectrum: pump intensity and phase-mismatch of the interacting beams. These two parameters can be used to optimise the small signal gain of an OPA. For the case of ideal phase-matching ($\Delta k = 0$), the gain becomes maximal $g = \Gamma$. But when $\Delta k \neq 0$, the shape of the small signal gain is defined by the interplay between the pump intensity and phase-mismatch. Equation (2.28) suggests that by increasing the pump intensity one could tailor the spectral shape of the amplification gain and minimize the effect of Δk and the resulting accumulated phase-mismatching, ΔkL.

To study this effect, numerical simulations (using SISYFOS code, for details see Appendix B) for OPCPA at two different pump intensities (50 and 100 GW/cm^2) and two different crystals with different phase-mismatch (BBO and LBO) are performed. In these simulations it is assumed that the 1 mJ pump pulses, centered at 515 nm, have a Gaussian distribution in time and space with a pulse duration of 1 ps at FWHM.

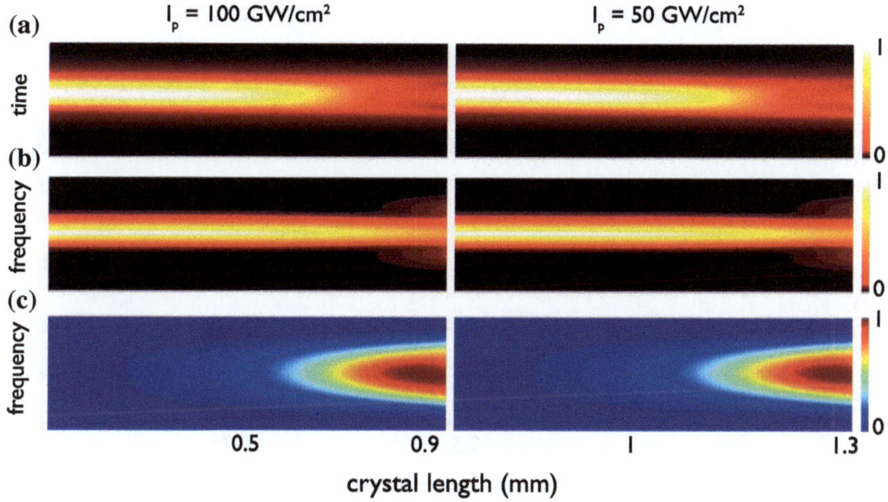

Fig. 2.3 Intensity-dependent gain in BBO. **a** The temporal profile of pump pulses over crystal length at two different pump intensities, and **b** their corresponding spectra. **c** Amplified signal spectrum over crystal length for BBO crystal used as a nonlinear medium for a saturated OPCPA

The signal pulse energy is set to 3 μJ. In each simulation, the crystal thickness is optimized to reach energy saturation at each intensity.

Figure 2.3 shows the temporal and spectral evolution of the pump pulses and the spectral evolution of the signal pulses over the length of a BBO crystal. The amplification for pump intensities of 100 and 50 GW/cm^2 reaches saturation in 0.9 mm-thick and 1.3 mm-thick BBO crystals, respectively. For both intensities the pump energy is depleted, which results in spectral broadening of the pump pulses Fig. 2.3b. The gain curves and therefore the amplified spectra for both cases are nearly identical. Figure 2.5b shows the accumulated phase-mismatch (ΔkL) for amplification in these two cases.

Due to the lower nonlinearity of LBO, a longer crystal length is required to reach saturation: 5.4 mm for the pump intensity of 50 GW/cm^2 and 2.5 mm for 100 GW/cm^2. Figure 2.4 shows that the saturation dynamics are different for the case of an LBO crystal. Here, for the pump intensity of 50 GW/cm^2 the pump energy is not depleted, although the amplification reaches saturation. As the signal is linearly chirped, and therefore the frequencies which satisfy phase-matching have temporal overlap with the wings of the pump pulses, mostly the energy of the temporal wings of the pump is extracted. Little spectral broadening in the pump spectrum is observed Fig. 2.4b and the amplified spectrum has a hole in the middle, which matches very well with the phase-mismatch curve shown in Fig. 2.5a. At 100 GW/cm^2 pump intensity the situation becomes similar to the BBO crystal. Despite the large phase-mismatch in the central frequencies of the spectrum, due to the effect of the higher pump intensity on the gain bandwidth, no hole is present.

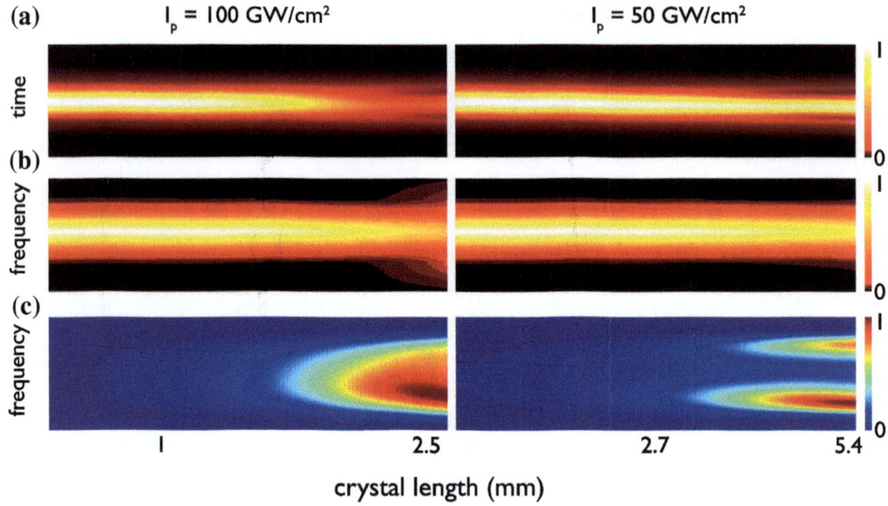

Fig. 2.4 Intensity dependent gain in LBO. **a** The temporal profile of pump pulses over crystal length at two different pump intensities, and **b** their corresponding spectra. **c** Amplified signal spectrum over crystal length for LBO crystal used as a nonlinear medium for a saturated OPCPA

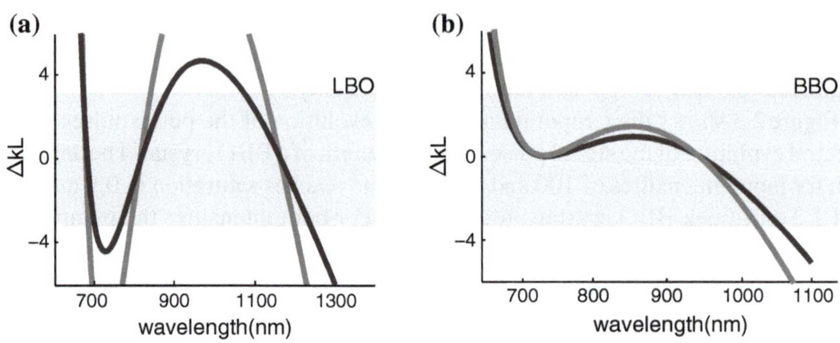

Fig. 2.5 **a** Residual phase-mismatch of a saturated OPCPA with a pump intensity of $100\,\mathrm{GW/cm^2}$ (*blue*) and $50\,\mathrm{GW/cm^2}$ (*red*) utilizing a 2.5 and 5.4 mm LBO crystal for the nonlinear medium and, **b** using a 0.9 and 1.5 mm BBO crystal as the nonlinear medium. The *pink area* shows the phase-mismatch between $-\pi$ and π

Therefore it can be concluded that high peak intensity of pump pulses, achievable in short-pulse-pumped OPCPA systems, ensures high amplification gain which results in saturation in a shorter crystal length. Additionally, as was shown above, it is a parameter for tailoring the amplification gain.

Remark Equation (2.28) indicates that beside the phase-mismatch between the interacting beams, pump intensity has an influence on the small signal gain. Increasing the pump intensity leads to higher value for small signal gain and compensate for a degree of phase-mismatch, allowing gain at wavelengths that are not amplified at lower intensity. This shows one of the advantages of short-pule-pumped OPCPAs, as higher pump intensity can be used without damaging the crystal.

Amplification and Saturation

In the large gain approximation ($\Gamma L \gg 1$), Eq. (2.27) simplifies to:

$$I_s(L) \cong \frac{1}{4} I_{s0} \exp(2\Gamma L). \tag{2.29}$$

$$G = \frac{I_s(L)}{I_{s0}} = \frac{1}{4} \exp(2\Gamma L). \tag{2.30}$$

The exponential growth in OPA shown in Eq. (2.30), is qualitatively different from the quadratic growth of other second order nonlinear phenomena. In an OPA, an initial signal photon stimulates the generation of an additional signal and an idler photon. Likewise, due to the symmetry of the signal and idler, the generated idler photon stimulates the generation of a signal photon. Therefore the generation of a signal field reinforces the generation of an idler field and viceversa. This generates the positive feedback responsible for the exponential growth of the signal.

The OPA amplification gain can be divided to three different regions Fig. 2.6. Region A, with exponential behavior, for $E_i \ll E_p$, where E_i and E_p are idler and pump electric fields, respectively. Region B, linear amplification until $E_i \approx E_p$ and region C, where $E_i \gg E_p$ and the direction of energy flow is reversed. Here the backconversion of energy from the idler and signal to the pump (sum frequency generation of idler and signal) overtakes the amplification process. Due to the symmetry in conservation of energy and momentum, energy flow from pump photons to signal photons and from signal photons to the pump photons happens simultaneously. But the backconversion of energy becomes noticeable when the

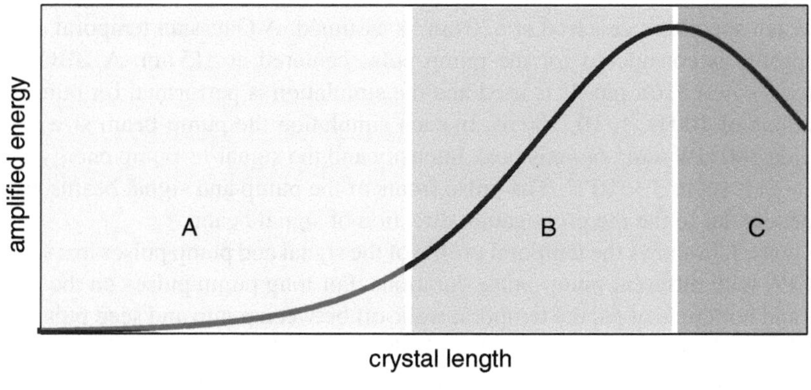

Fig. 2.6 Typical OPA gain curve as a function of crystal thickness. For the description of the regions A, B and C, see Sect. 2.1

total energy of idler and signal pulses becomes comparable to or more than the pump energy. In an adequately long crystal operating beyond region *C*, the energy between pump and signal is shuffled repeatedly back and forth.

Remark The generation of the idler is the initial fuel for the exponential raise in the gain of parametric amplification, as is shown in Fig. 2.6—region A. In region B, the behavior of the gain becomes linear. Therefore the elimination and control of the idler would lead to better energy extraction by the signal pulses and consequently higher efficiency. This is the basics for the efficient OPCPA systems, which will be discussed in Chap. 5.

Optimum Pump-Pulse Duration

In an OPCPA, the pulse duration of the pump laser has an important role in the amplification process. On one hand, the damage threshold intensity for transparent materials increases roughly proportional to the inverse of the square root of the pulse duration ($I_{th} \propto \tau^{-\frac{1}{2}}$) for pulsed lasers, following the dynamics of impact ionization. For lasers with pulses longer than µs, thermal damage mechanisms may dominate, but such laser sources are not considered here (Stuart et al. 1996).

On the other hand, the impact of temporal walk-off between pump and seed pulses increases with decreasing pulse duration. These two effects place opposing demands on the pulse duration, suggesting that an optimal compromise exists. To study this, a series of simulations were performed to find the optimum pump-pulse duration for broadband OPCPAs by varying the pump-pulse duration from 100 fs to 100 ps. In the first series, the peak intensity on the crystal is set to $100\,\mathrm{GW/cm^2}$ for all pump-pulse durations in order to decouple the effect of the following parameters: intensity-dependent gain, temporal walk-off between signal and pump pulses and the damage threshold intensity. Here, the shape of the amplified spectrum is affected by the temporal walk-off between pump and seed pulses and the influence of the intensity-dependent gain is decoupled by using a similar intensity for all pump-pulse durations.

In the following simulations, a linearly chirped signal with a fourth-order super-Gaussian spectrum, centered at 820 nm, is assumed. A Gaussian temporal and spatial profile is considered for the pump pulse centered at 515 nm. A BBO crystal with $d_{eff} = 2.3168\,\mathrm{pm/V}$ is used and the simulation is performed for pump-pulse durations of 100 fs, 1, 10, 100 ps. In each simulation the pump-beam size is set to provide $100\,\mathrm{GW/cm^2}$ on-axis peak intensity and the signal-to-pump energy ratio in all cases is set to 3×10^{-3}. The pulse fronts of the pump and signal beams are both perpendicular to the the propagation direction of signal beam.

Figure 2.7a shows the temporal profile of the signal and pump pulses in a saturated OPCPA with different pump-pulse durations. For long pump pulses on the order of tens and hundreds of ps, the temporal walk-off between pump and seed pulses in the OPCPA nonlinear crystal is negligible. In the saturated regime the backconversion of energy is visible in the center of the pulse, due to the Gaussian temporal profile of the pump. For 1 ps pump pulses, the temporal walk-off between the interacting beams becomes visible and finally, for 100 fs pump pulses, the temporal walk-off limits the

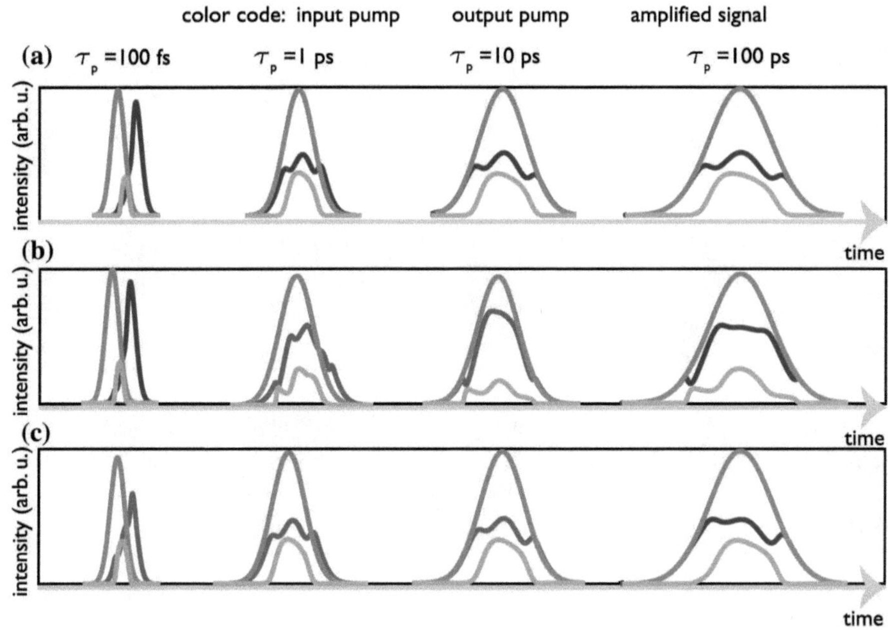

Fig. 2.7 Temporal profile of pump and signal pulses in a **a** saturated and **b** over-saturated OPCPA for different input pump-pulse durations. The temporal walk-off for long pulses is negligible. **c** Saturated OPCPA in which the pump intensity is scaled relative to the pump-pulse duration. *Green* pump temporal profile before amplification, *blue* residual pump temporal profile after amplification, *red* amplified signal temporal profile

efficient energy extraction from the pump, as the signal and pump pulses quickly become temporally separated before backconversion of the energy to the pump pulse starts. This phenomenon causes local saturation points in the gain of the OPCPA.

Figure 2.7b shows the behavior of these cases in an over-saturated OPCPA, where a noticeable amount of amplified signal energy goes back to the pump pulse, causing modulations on the temporal profile of the amplified signal.

This line of reasoning also applies to the spatial profile of the pump pulses. Figure 2.8a shows the effect of saturation in space. Poor energy extraction in 100 fs pump pulses leaves the spatial shape of the pump almost unchanged, whereas for longer pump pulse durations the amplification reaches saturation and the energy backconversion is visible in the center of the beam, which carries the highest intensity. In the case of an over-saturated OPCPA (Fig. 2.8b) the residual pump beam profile undergoes severe changes with its inverse imprinted on the amplified signal beam profile (Fig. 2.9b), except for 100 fs pump pulses, due to localized saturation.

As can be seen in Fig. 2.10b pump pulse durations longer than 1 ps at the same peak intensity result in approximately the same amplified signal peak power while the amplified peak power drops for 100 fs pump pulses. This indicates the negligible

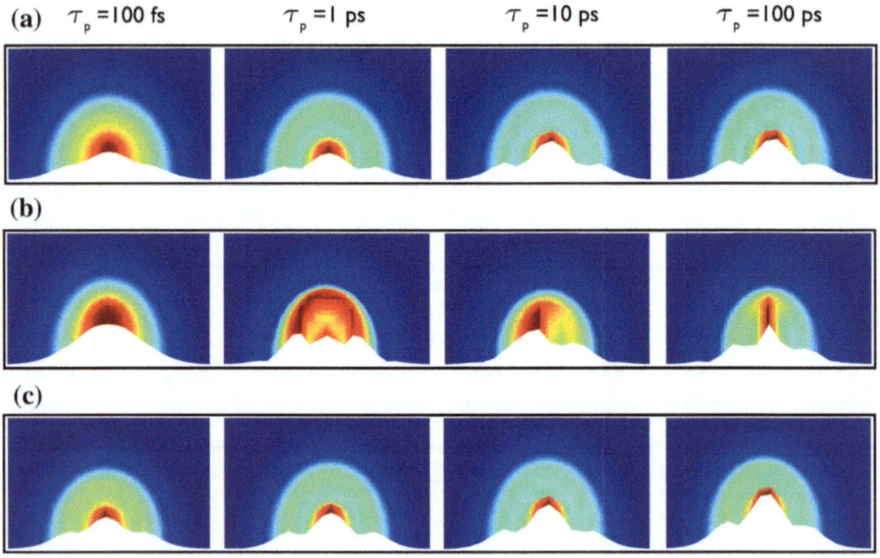

Fig. 2.8 Spatial profile of pump beam after amplification in a **a** saturated and **b** over-saturated OPCPA for different pump-pulse durations. The effect of spatial walk-off for 10 ps pump pulses which leads to asymmetric energy extraction of the pump is clear. **c** Saturated OPCPA in which the pump intensity is scaled relative to the pump-pulse duration. The *white areas* shows the on-axis beam profile

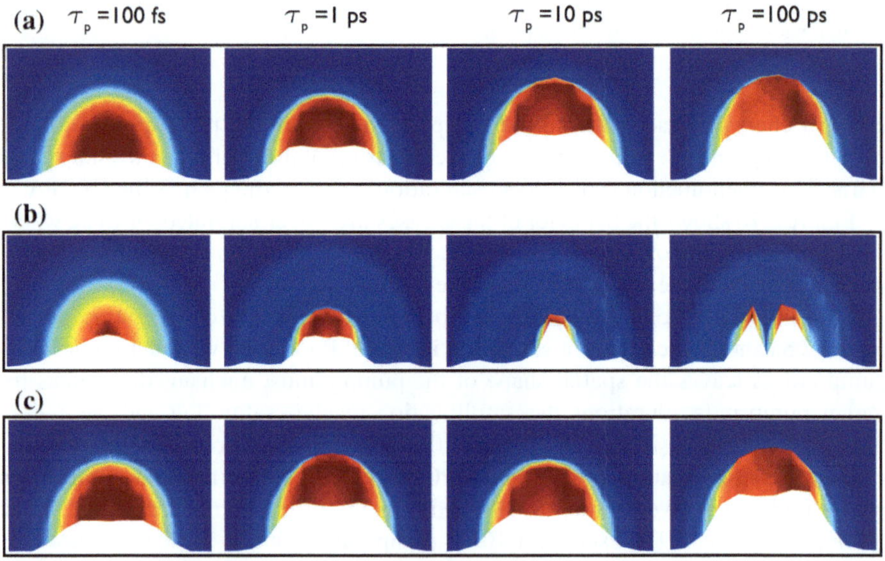

Fig. 2.9 Spatial profile of seed beam after amplification in a **a** saturated and **b** over-saturated OPCPA for different pump-pulse duration. **c** Saturated OPCPA in which the pump intensity is scaled relative to the pump pulse-duration. The *white areas* show the on-axis beam profile

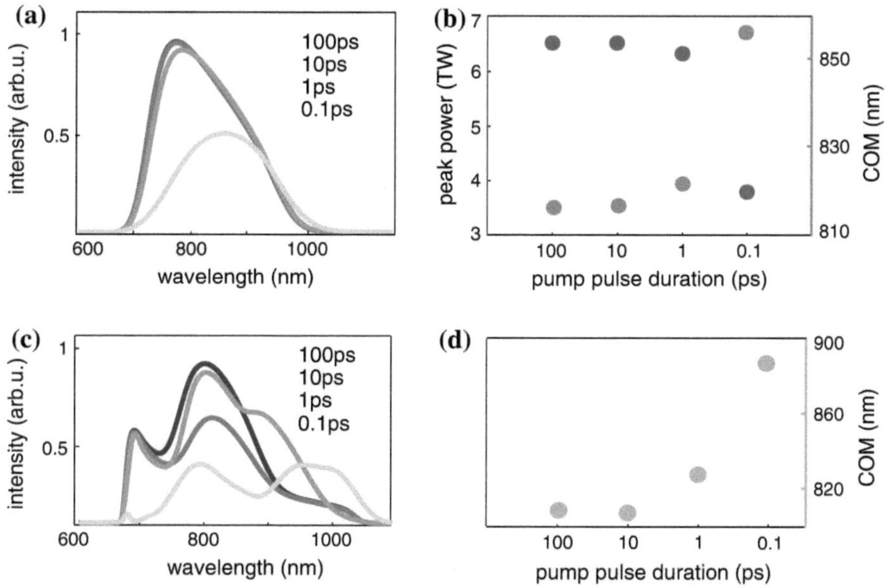

Fig. 2.10 **a** Amplified spectra for different pump-pulse durations in a saturated OPCPA. The amplified spectra of 10 and 100 ps pump pulses are overlapped. **b** Peak power of the amplified signal energy and the corresponding spectral center of mass (COM) in a saturated OPCPA versus pump-pulse duration. **c** Amplified spectra for different pump-pulse durations and, **d** peak powers of the amplified signal energy and the corresponding spectral center of mass (COM) in an over-saturated OPCPA

effect of temporal walk-off for pump pulses longer than 1 ps and the advantage of near-1-ps-pumped OPCPAs.

As the signal has a linear positive chirp, the center of mass of the amplified spectrum moves to longer wavelengths for shorter pump-pulse durations, because the leading edge of the signal pulse always overlaps the undepleted pump pulse regions due to the temporal walk-off between pump and signal (Fig. 2.10a, b). In an over-saturated OPCPA, the amplified signal spectrum becomes modulated and broader. Here, imperfectly phase-matched spectral components of the signal spectrum have a chance to be amplified, because of energy backconversion of the phase-matched spectral components and result in a broader amplified spectrum. The longer nonlinear crystal used for the over-saturated OPCPA causes a larger temporal walk-off and shifts the spectral center of mass to even longer wavelengths compared to the saturated OPCPA (Fig. 2.10c, d).

In the second series of simulations, all of the input parameters are the same as the previous simulations except for the on-axis pump intensity in the nonlinear crystal. Here the pump intensity is scaled relative to the pump-pulse duration to stay below the damage threshold intensity. The damage threshold intensity for 1 ps pulses is assumed to be 100 GW/cm^2 based on experimental measurements and the relative intensity is scaled with the square root of the pump-pulse duration. As the damage

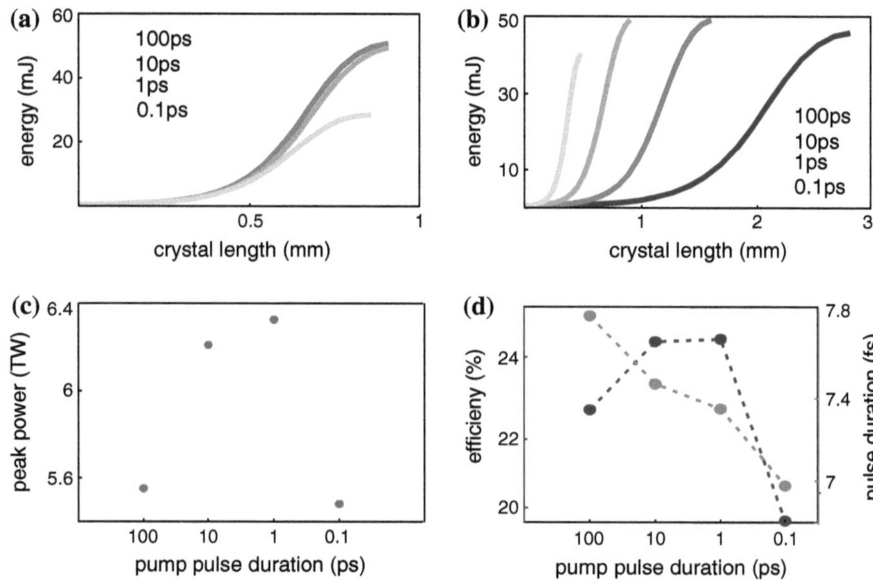

Fig. 2.11 **a** Amplified signal energy over crystal length for different pump-pulse durations with a fixed peak intensity of 100 GW/cm^2. The energy curves for 100 and 10 ps overlaps. **b** Amplified signal-energy scaling with nonlinear crystal length for different pulse durations and with the pump intensity scaled relative to the pump-pulse duration. **c** Peak power of the amplified signal. **d** Pump-to-signal conversion efficiency and the amplified signal-pulse duration of OPCPAs with different pump pulse durations

threshold intensity is higher for short pump pulses, the energy extraction of pump pulses becomes more efficient (Figs. 2.7c and 2.8c), and for different pump-pulse durations, saturation occurs at different crystal thicknesses (Fig. 2.11). Therefore, for a longer pump pulse duration a thicker nonlinear crystal is needed to reach saturation. This limits the bandwidth of the amplified spectrum. The combination of all these parameters indicates that near-1-ps pump-pulse duration is the optimum for achieving the highest peak power in an OPCPA (Fig. 2.11).

Conclusion Pump pulse durations in the range of 1–10 ps are optimum for OPCPA systems in terms of spectral bandwidth and conversion efficiency. In this range the temporal walk-off between the interacting beam is negligible, the intensity damage threshold of materials are relatively high and thin crystals with high single-pass gain can be employed allowing large amplification bandwidth.

Yb-doped amplifiers, in thin-disk, slab, and bulk geometry provide such high energy pulses, making them suitable for pumping high-energy, high-power OPCPA systems.

Optimum Pump-to-Signal Energy Ratio

To understand efficiency and gain, let us assume an oak tree with 100 acorns. For a small number of hungry squirrels, there are more than enough acorns to go around and each squirrel has a chance to gather as many as he or she likes, or scientifically speaking, each squirrel has a high gain. But, because there are far fewer squirrels than acorns, their efficiency in picking up the acorns is low. Increasing the number of squirrels will increase the thoroughness of the consumption of acorns but at the same time decrease the share of acorns for each squirrel. A similar phenomenon happens in an OPA, by considering signal photons as squirrels and pump photons as acorns. By increasing the signal-to-pump energy ratio, the energy extraction and OPA efficiency increases.

The process can be explained by looking at the role of the idler. For a high input signal energy, saturation occurs in a shorter crystal length. As a result, the newly generated idler has a broader spectral bandwidth. These new photons generated at the edge of the idler spectrum correspond to amplification of seed photons that would not get amplified in a longer crystal, and result in a higher net efficiency. Essentially, the increased gain bandwidth allows for a seed pulse with the same stretched duration, normalized to the pump duration, to make use of a wider temporal window of pump pulse energy. For a low input-signal energy, energy saturation happens in a longer crystal length, which limits the bandwidth of the generated idler and therefore the energy-extraction efficiency.

For signal-pulse energies higher than the pump-pulse energy, although unfamiliar to our common sense, amplification still happens due to the conservation of momentum and energy in the generation of the idler. Figure 2.12 shows the calculated gain and efficiency for different signal-to-pump energy ratios. A signal-pump energy ratio of the order of 10^{-3} is the optimum for having a high amplification gain and efficiency simultaneously.

The signal-to-pump energy ratio also has an effect on the saturation behavior of an OPCPA. Figure 2.13 (right column) shows the amplification gain for different input signal energy. When the input signal is equal to the input pump energy, the amplified signal reaches saturation very quickly, 46 % of the amplified energy converts back to the pump and the process reaches a steady state for longer crystal lengths. By decreasing the input signal energy, the ratio of the back-converted energy decreases. For lower input signal energy, the steady-state is not observed and in addition to the first saturation point, a second one with higher energy appears (for energy ratio = 0.0001). The local saturation points are due to the fact that different spectral components reach saturation at different crystal lengths Fig. 2.13 (left column). For high signal energies, all of the frequencies reach saturation in a very short length of the crystal and by decreasing the signal energy, the saturation points spread over a larger range of crystal lengths.

Spectral components at 720, 790, 860, 930, 1000, and 1070 nm are considered for detailed study. Figure 2.14a shows the difference between the crystal length where the first wavelength and the last wavelength reach saturation (ΔL_{sat}), divided by the saturation length (L_{sat}) of the whole spectrum in the crystal. Figure 2.14b, c shows the relative saturation length for different wavelengths and input signal energies and their corresponding amplification gain, alternatively. Saturation is reached quickest at 870 nm due to the small phase-mismatch. For lower signal energies and longer crystal lengths, the saturation lengths for different wavelengths become more dispersed.

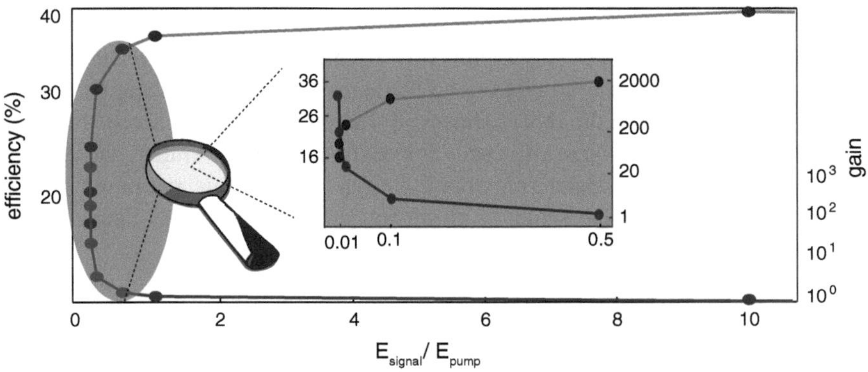

ratio	10^1	1	0.5	10^{-1}	10^{-2}	10^{-3}	10^{-4}
gain	1.04	1.4	1.7	4.1	26	201	1661
efficiency 1 (%)	40	37	36	31	25	20	17
efficiency 2 (%)	64	60	57	50	40	32	27

Fig. 2.12 Calculated gain and efficiency for different signal-pump energy ratios in an OPCPA pumped with 1 mJ pulses at 515 nm. The amplified spectrum in a BBO crystal covers 680–1050 nm. The amplification gain is calculated by dividing the amplified signal energy over the input signal energy. Efficiency 1 is the net increase in signal energy divided by the input pump energy and efficiency 2 is the sum of the net increase in signal energy plus the idler energy divided by the input pump energy

Conclusion The signal-to-pump energy ratio is a crucial parameter in an OPCPA, as it directly affects the amplification gain, conversion efficiency and the level of superfluorescence. Based on the analytical study in this section, it is shown that this parameter influences the gain behavior of broadband OPCPA dramatically. In high-seeded OPCPA, saturation of all the signal frequencies occur simultaneously whereas for for the case of low-seeded OPCPA the saturation of different frequencies is spread over a larger length of the crystal. This modifies the shape of the amplified spectrum and limits the efficient energy extraction from the pump pulses.

It is shown analytically that the signal-to-pump energy ratio of 10^{-3} results in a simultaneous high amplification gain and high conversion efficiency.

OPA Phase

In the OPA process, due to the accumulated phase slip $|\Delta k L|$, caused by the phase-mismatch between the interacting waves, an additional phase is imprinted on the signal wave during amplification (Herrmann et al. 2010c). In the case of undepleted pump, the so called OPA phase is defined as:

$$\Phi_s(z = L) = \Phi_{s(0)} - \frac{\Delta k L}{2} + \arctan \frac{\Delta k \times \tanh[L(g^2 - (\Delta k/2)^2)^{1/2}]}{2(g^2 - (\Delta k/2)^2)^{1/2}}, \quad (2.31)$$

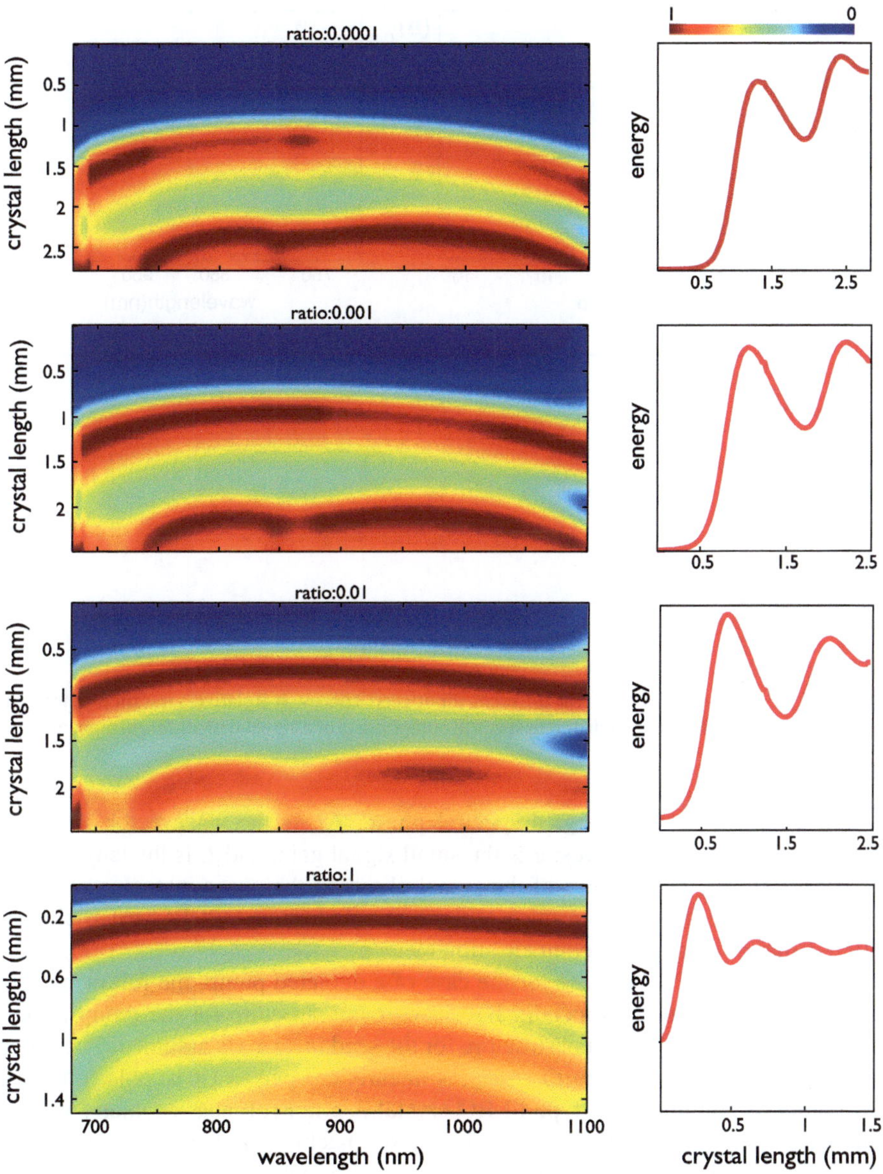

Fig. 2.13 Spectral behavior of an amplified signal for different seed-to-pump energy ratios. *Left* spectral distribution of the amplified signal at each position in the crystal. The amplitude of each spectral component over the crystal length is normalized to one. *Right* the integrated spectrum at each crystal position

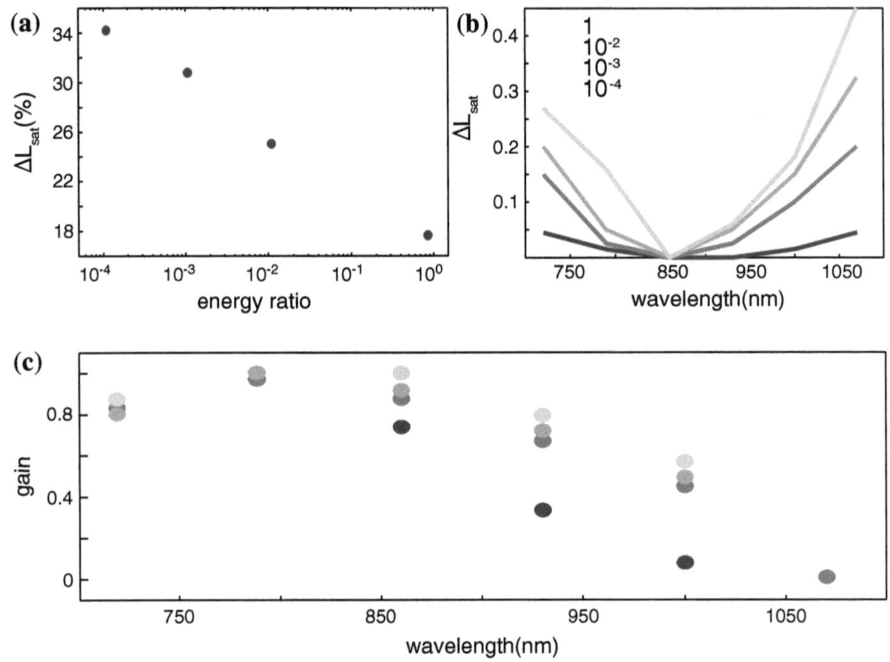

Fig. 2.14 **a** $\Delta L_{sat} \times 100/L_{sat}$ for different input seed energy. **b** Relative saturation length and **c** relative amplification gain for different spectral components of a signal with different input energies

where $\Phi_{s(0)}$ is the initial signal spectral phase, Δk is the wavevector mismatch between the interacting waves, g is the small signal gain, and L is the length of the OPA crystal. The dependence of the signal phase on the pump intensity, shown in Eq. (2.31), makes the compensation of the spectral phase in OPCPA very tricky and often demands an adaptive compensation element.

To study the OPCPA phase for the case of a depleted pump, the following simulations are performed (Appendix B). A Gaussian structure for the pump in both space and time and a Gaussian structure for the seed in space is assumed. The super-Gaussian seed spectrum (fourth order), is linearly chirped to 1.6 ps pulse duration. The signal is amplified in a type I LBO crystal in a noncollinear geometry with an internal pump-signal angle of 1.1° and a phase-matching angle of 14.4°, assuming the d_{eff} value of 0.82 pm/V. The initial pump and signal energies are set to 3 mJ and 3 μJ, respectively. The beam size of the pump is adjusted to provide 100 GW/cm^2 peak intensity on the crystal. The simulation is performed for different amounts of gain by changing the crystal thickness.

The OPCPA reaches saturation in a 2.5 mm thick crystal. To extract the OPCPA phase, the beam after amplification is focused analytically using a convex Bk7-lens with a radius of curvature of 0.6 m. The spectral phase of the central point of the focused beam is then calculated and subtracted from the spectral phase of the input signal. Therefore the calculated phase just includes the OPCPA phase and the introduced chirp on the pulse due to the dispersion of LBO. Later this dispersion is compensated numerically.

Figure 2.15 shows the residual higher order OPCPA phase for different gain factors and the corresponding pulses in the time domain. For low gain, the shown third order OPCPA phase follows the phase-mismatch of the interacting waves in the LBO crystal (as shown in Fig. 2.5). In a saturated and an over-saturated OPCPA (green and blue curves in Fig. 2.15a, respectively) the residual phase, mainly third order, increases and results in pronounced satellite pulses in the time domain. Therefore, to compensate the chirp of amplified pulses in an OPCPA system, the OPCPA phase should be carefully measured and included in the design of the compressor.

Conclusion The analytical study on the behavior of the imposed spectral phase on a signal pulse during amplification is presented in this section.

The results indicate that in addition to the spectral phase caused by dispersion in the OPCPA nonlinear crystal, the accumulated phase-mismatch between the interacting beams results in a higher order dispersion (OPA phase). This higher order dispersion increases when the amplification is saturated due to the additional effect of the temporally modified pump pulses in this regime.

The intensity dependence of the OPCPA phase suggests the correlation of the wavefront of the amplified beam to intensity. Therefore in the amplified beam, each frequency may have a different divergence.

2.2 Yb:YAG Based OPCPA Systems

Considering the concept of OPCPA explained in the previous sections, different low-order harmonics of high-energy narrowband lasers could be utilized to amplify a broad range of frequencies extending from ultraviolet to infrared. This allows a variety of operational modes, offering multicycle pulses tunable over several octaves, few-cycle pulses at different carrier wavelengths, and multi-octave synthesis of light waveforms.

Diode-pumped ytterbium-doped (Yb-doped) thin-disk lasers are reliable, cost-effective, and power-scalable sources of high-energy 1-ps-scale laser pulses. These characteristics make them suitable sources for pumping OPCPA systems. In what follows, the feasibility of a few-cycle OPCPA and a tunable OPCPA based on diode-pumped Yb-doped thin-disk lasers are discussed, while Chaps. 5 and 6 cover the detailed study of few-cycle OPCPA and subcycle light waveforms.

Fundamental-Pumped OPCPA Three types of phase-matching schemes can be used to amplify a broadband spectrum supporting ultra-short intense pulses, utilizing near-1-ps pulses centered at the fundamental wavelength of 1030 nm: collinear, noncollinear and quasi phase-matched.

Periodically poled $LiNbO_3$ (PPLN) and periodically poled stoichiometric $LiTaO_3$ (PPSLT) are suitable crystals for few-cycle OPCPA systems in the MIR range, as they have a very high nonlinearity, e.g. $d_{33} = 25\,\mathrm{pm/V}$ for $LiNbO_3$, resulting in a

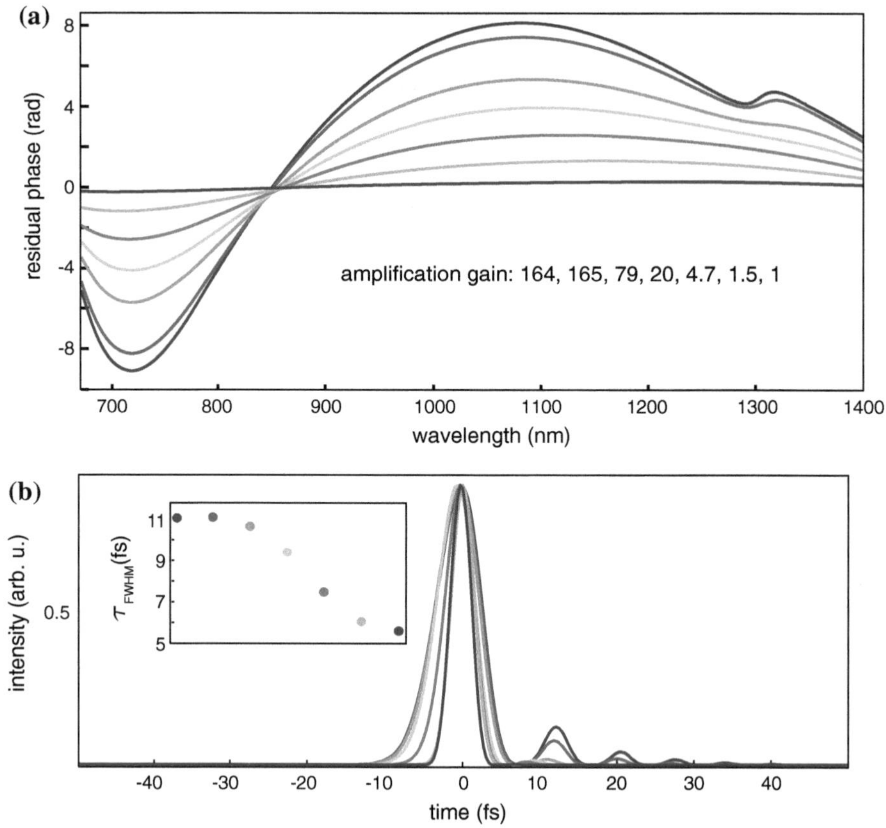

Fig. 2.15 **a** The residual spectral phase in OPCPAs with different gain after compensation of the linear dispersion of the amplifier and **b** their corresponding pulses in the time domain. *Inset* the pulse duration at FWHM

high amplification gain and therefore efficiency. However the limited aperture size of the periodically poled crystals limits their usability to low-energy-pumped OPCPA systems or pre-amplification stages to boost the signal energy-level before the high-energy OPCPA stages. Alternatively, bulk crystals like $LiNbO_3$, $LiIO_3$ or BBO can be used for amplification.

Phase-matching in a BBO crystal can be achieved in a noncollinear type-I geometry with an internal pump-signal angle of 2.2° and a phase-matching angle of 20°. Although amplification in a BBO crystal with a length of 2 mm supports pulses as short as 19 fs with a carrier wavelength of 2 μm, the linear absorption coefficient of the crystal at 2.55 μm reaches $0.5\,\text{cm}^{-1}$ and the accumulated heat in this process may impede the OPCPA conversion efficiency or cause crystal damage. For an internal pump-signal angle of 3°, the center of the amplified spectrum moves to higher frequencies, but the amplified spectral bandwidth only supports 43 fs pulses (Fig. 2.16).

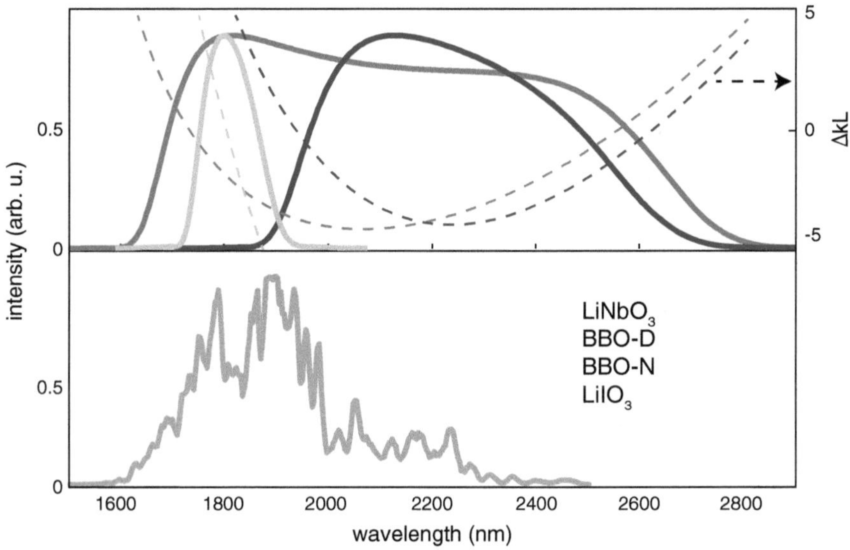

Fig. 2.16 *Top* Amplified spectrum in a 4 mm $LiNbO_3$, a 4 mm BBO crystal in a collinear geometry (BBO-D), and a 2.1 mm BBO crystal in a noncollinear geometry (BBO-N). In the mentioned simulations, the pump energy of a 1.6 ps pump pulses, centered at 1030 nm, with the Gaussian profile in time and space is set to 1 mJ. The stretched broadband seed pulses have a duration of 1.6 ps, a Gaussian profile in time, and contain 0.5 μJ of energy. The *dashed lines* show the accumulated OPA phase over 2 mm-thick crystals. *Bottom* the amplified spectrum in a 2 mm thick $LiIO_3$ crystal in the first stage of the OPCPA system described in (Deng et al. 2012)

On the other hand, using a type I $LiNbO_3$ crystal with a phase-matching angle of 47° at the degeneracy point in a collinear geometry or a type I $LiIO_3$ supports few-cycle pulses (Fig. 2.16), but $LiIO_3$ suffers from a low damage threshold intensity.

In reality, even in a "collinear" degenerate OPCPA a small angle between the pump and seed beams is required to spatially separate the signal and the idler in order to avoid interference between them. This separation is necessary when the carrier envelope phase (CEP) offset of signal pulses are stable, as the generated idler pulses are not CEP stable. Even for an OPCPA with a CEP stabilized pump and signal, which results in a CEP stabilized idler, a small angle is required for pulse compression as the idler and signal possess opposite chirp. Figure 2.17 shows the effect of the noncollinear angle on the shape of the amplified spectrum and the OPCPA gain in a 4 mm-thick $LiNbO_3$ crystal. The simulation is performed for a pump energy of 1 mJ at 1030 nm and a signal input energy of 0.5 μJ.

Among the mentioned crystals, $LiNbO_3$ is the most suitable one for high-energy OPCPA systems as it supports a broad amplification bandwidth, is available with large apertures, and has a high d_{eff} value of 3.3 pm/V where d_{31}= 4.35 pm/V; although its n_2 value is three times higher than BBO and the applicable pump intensity in the crystal is limited to a value below 70 GW/cm^2 due to photorefraction (Furukawa 2000; Augstov and Shvarts 1980).

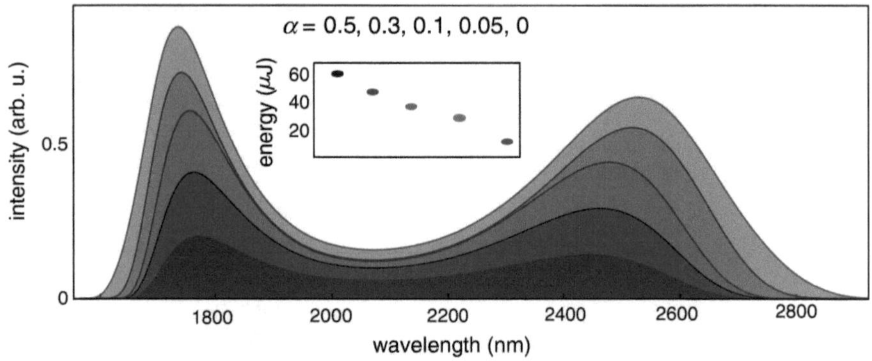

Fig. 2.17 The effect of the noncollinear angle (α) on the amplified spectrum in a 4 mm $LiNbO_3$ crystal. *Inset* the corresponding effect of the noncollinear angle on the amplification gain. In this graph, the relative amplitude of the spectra does not show their relative energy

Second-Harmonic Pumped OPCPA By using the frequency doubled output of Yb-doped lasers to pump an OPCPA, the amplification gain can be shifted to the NIR range of the spectrum, extending from about 650–1400 nm. As the zero-dispersion point of the majority of nonlinear crystals is above 1300 nm, broadband-collinear-degenerate OPCPA is not an option. Therefore, to achieve a broadband amplification, noncollinear geometry should be used.

Crystals like YCOB (Pires et al. 2014), BBO (Bromage et al. 2011; Rothhardt et al. 2012; Schultze et al. 2010), LBO (Thai et al. 2010) and DKDP (Skrobol et al. 2012), pumped at 515 nm, support broadband phase-matching in a noncollinear geometry. Among these crystals, LBO and BBO are distinguished by their high nonlinearities and broadband amplification gain spectra.

BBO is a robust, negative uniaxial crystal but limited in available aperture size to $25 \times 25\,\text{mm}^2$. LBO is a biaxial crystal supporting broad phase-matching in the XY plane, where $\theta = 90°$ (Zhao et al. 2008). The crystal in this plane can be treated similarly to a negative uniaxial crystal with $n_o = n_z$ and

$$\frac{1}{\lambda_p} = \frac{1}{\lambda_s} + \frac{1}{\lambda_i}, \tag{2.32}$$

$$(\frac{n_i}{\lambda_i})^2 = (\frac{n_s}{\lambda_s})^2 + (\frac{n_p}{\lambda_p})^2 - 2\cos(\alpha)\frac{n_s n_p}{\lambda_s \lambda_p}, \tag{2.33}$$

$$\frac{1}{n_p^2} = \frac{\cos^2(\theta)}{n_{py}^2} + \frac{\sin^2(\theta)}{n_{px}^2}. \tag{2.34}$$

One of the advantages of the noncollinear geometry is the possibility of adjusting the amplified spectrum by changing the phase-matching angle and the noncollinear angle. Here, by changing the phase-matching angle θ, the amplification bandwidth

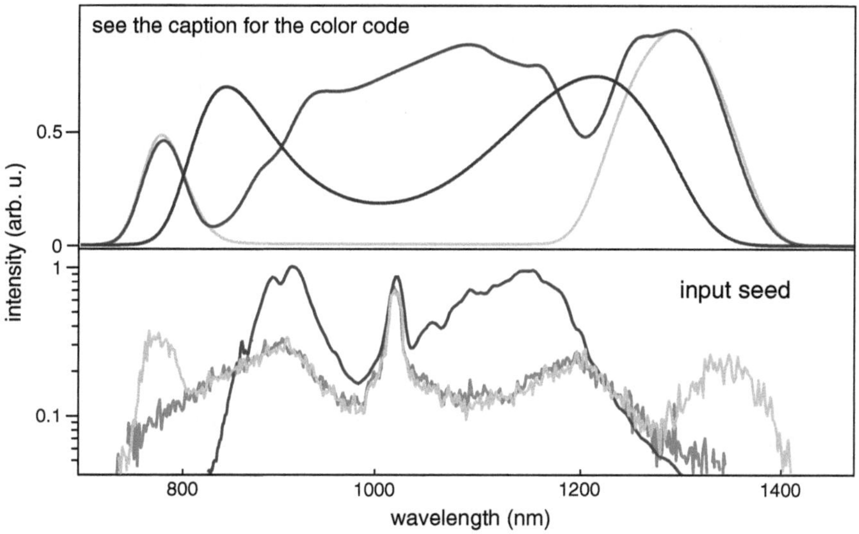

Fig. 2.18 *Top* simulated amplified spectra in a 2-stage OPCPA employing LBO crystals with different phase-matching angles. In the first OPCPA stage, the phase-matching angle of θ : 13.8° is used to amplify the input spectrum in its wings (*light blue line*). In the second stage the phase matching is tuned to θ : 14.9°, to amplify the middle of the input spectrum (*dark blue line*). The *black line* shows the amplified spectrum in a 2-stage OPCPA system using LBO crystals with a phase-matching angle of 14.4°, for comparison. It can be seen in the first case broader spectrum is amplified. *Bottom* Experimental proof of the mentioned simulation. Here by using a 4 mm LBO crystal and for different phase-matching angles, the wings or the middle of the input spectrum is amplified. The OPCPA seed in this measurement was generated in a 4 mm TGG crystal. The supercontinuum is passed through a 95 % output coupler to suppress the intense spectral components in the vicinity of 1030 nm (Appendix D)

can be tuned around a fixed central frequency. Figure 2.18 shows the simulated amplified spectrum in a 2-stage OPCPA system. The simulation is performed for a Gaussian pump pulse in time and space with an energy of 1 mJ and a pulse duration at FWHM of 1.6 ps while the beam size is set to 740 μm at FWHM. The input seed energy is 1 μJ.

In the first amplification stage, with a phase-matching angle of 13.8° and a non-collinear angle of 1.1°, the amplification gain is moved to the wings of the spectrum. Changing the phase-matching angle of the second stage to 14.9° shifts the gain to the center of the spectrum. This spectrum is compared with a 2-stage OPCPA system with LBO crystals at a phase-matching angle of 14.4°. As shown in Fig. 2.18, the first scheme results in a somewhat broader spectrum. Figure 2.18-bottom shows the experimental demonstration of this process in an OPCPA stage seeded with a supercontinuum generated in a TGG crystal (Appendix D).

This flexibility for the gain bandwidth engineering in NOPA systems is theoretically and experimentally investigated in Sect. 5.3 by utilizing BBO and LBO crystals.

material	LBO	BBO	KBBF	KDA	DKDP
d_{eff} (pm/V)	0.8	2	0.45	0.5	0.3

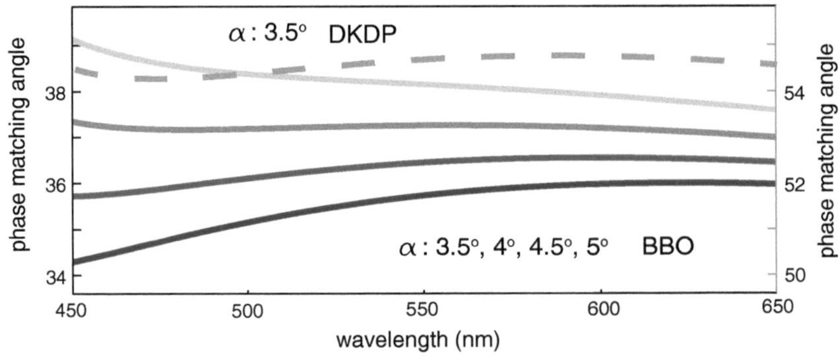

Fig. 2.19 *Top* comparison between the nonlinearity of different crystals suitable for third harmonic-pumped OPCPA. *Bottom* Phase-matching angles for broadband amplification in BBO and DKDP crystals pumped at 343 nm

Third-Harmonic Pumped OPCPA The third-harmonic of an Yb-doped laser can be obtained via sum frequency generation using the fundamental frequency and the second harmonic of the laser. The OPCPA crystal in this range should have a high band-gap energy to avoid resonant absorption in the OPCPA. Among different groups of nonlinear crystals, the borate group is the most suitable to be used in the ultraviolet region due to their high nonlinear coefficients, transparency and high laser induced damage thresholds. A broadband amplification gain in a noncollinear OPCPA, pumped at 343 nm, can be achieved in DKDP and BBO crystals (Fig. 2.19), but BBO is more favorable due to its higher d_{eff} value.

A serious challenge in short-pulse-pumped and short-wavelength-pumped OPCPAs are higher-order nonlinear effects in the wavelength range of the pump beam. As the pump wavelength gets closer to the resonance frequency of the crystal, $\chi^{(3)}$ grows rapidly and as a result, the efficiency of third order nonlinear effects like self-phase modulation (SPM) and third-harmonic generation (THG) in the crystal increases drastically. The third harmonic of the pump at 343 nm is at 114 nm. The linear absorption coefficient of BBO for 266 nm is 0.15 cm^{-1} and the crystal is not transparent for wavelengths below 190 nm. This absorption could lead to damage in the crystal.

The other issue in this range is two photon absorption (TPA). The reported TPA coefficient at 211 nm is 2.43 cm/GW for BBO and 1.03 cm/GW for LBO. These numbers decrease at 264 nm to 0.93 cm/GW for BBO and 0.15 cm/GW for LBO. Due to the UV cutoff edge at 189 nm for BBO and 160 nm for LBO, TPA is a severe energy loss mechanism for intense ultrashort UV pulses, which also contributes to pulse broadening (Dubietis et al. 2000).

A numerical simulation software SISYFOS (Appendix B), is used to estimate TPA in a 343 nm-pumped OPCPA. The matrix used to describe the third order effects consists of $[i \times nb + j]$ elements, which describes how beam j induces absorption of beam i, where nb is the number of beams. As there is hardly any measured data on non-degenerate TPA coefficients and the beam with the highest frequency has the maximum TPA, the TPA is only calculated for the pump beam. The TPA coefficient for a BBO crystal at 354.7 nm and a 17 ps pulse is 1×10^{-11} cm/W. Therefore, for the pump pulses with an intensity of $100\,\mathrm{GW/cm^2}$, the absorption coefficient becomes $1\,\mathrm{cm^{-1}}$.

The measured damage threshold intensity for a BBO crystal irradiated with several ns-pump pulses is $48\,\mathrm{GW/cm^2}$ at 532 nm, and $25\,\mathrm{GW/cm^2}$ at 355 nm (Nakatani et al. 1988), which is expected to increase for ps-pump-lasers. The estimated TPA, using the mentioned parameters, is negligible for 40 mJ, 1.6 ps pulses at 343 nm.

In an alternative approach, to generate a few-cycle pulse in the VIS, a thin BBO crystal can be used to generate the second harmonic of a broadband OPCPA signal, pumped by the second harmonic of the laser. The gain bandwidth of the second-harmonic signal is inversely proportional to the length of the crystal and the efficiency scales quadratically in the unsaturated regime. Figure 2.20 shows the simulated second harmonic spectrum of a 5 fs pulse with Gaussian distribution in time and space.

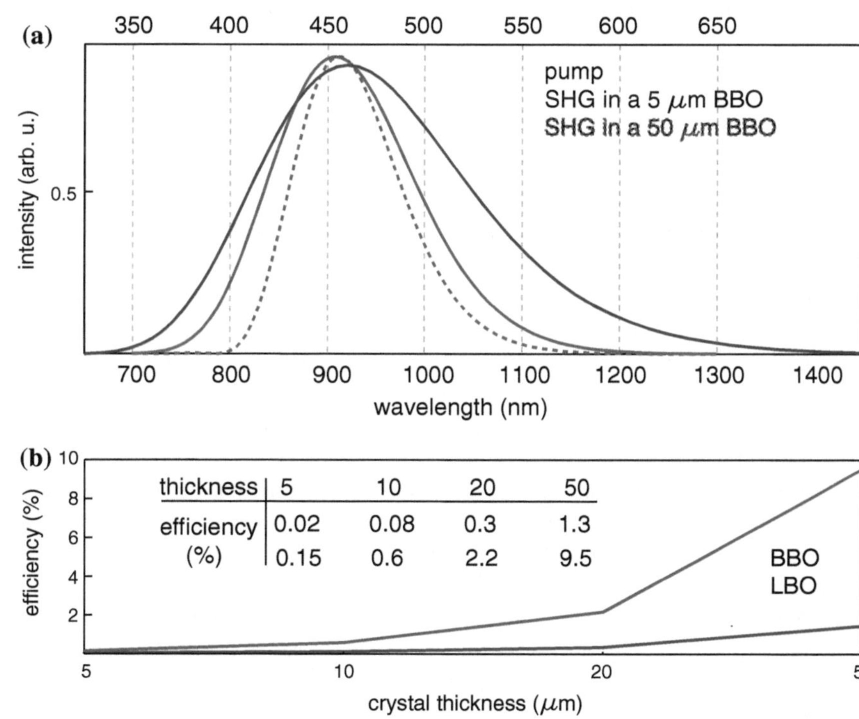

Fig. 2.20 **a** Simulated SHG in a 5 and 50 μm BBO crystal. **b** SHG efficiency for BBO and LBO crystals of different thicknesses

The beam size is set for 800 GW/cm^2 peak intensity on the crystal. The phase-matching angle of a type I BBO crystal is set to 25.6° and for a type I LBO it is 15.5°. The second-harmonic generation (SHG) efficiency in LBO is much lower than BBO for the same crystal thickness, but LBO supports a broader second harmonic spectrum. The same spectrum can be achieved for 50 μm-thick BBO with an efficiency of 9.5 % and 100 μm-thick LBO with an efficiency of 5 %. To overcome this poor conversion efficiency, achromatic SHG technique can be used. Here, the fundamental beam can be angularly dispersed to facilitate the phase-matching of a wide range of frequencies, which leads to efficiencies as high as 20 % (Wnuk et al. 2010). However, the technique requires a complicated setup to minimize the angular dispersion of broadband pulses. Therefore, OPCPA is a more promising route to achieve broadband, high energy pulses in the VIS range.

Conclusion By employing different harmonics of an Yb:YAG thin-disk laser to pump an OPCPA system, few-cycle, high-energy, pulses at different carrier wavelength from visible to mid-infrared can be generated. Among the variety of crystals and phase-matching techniques, degenerate OPCPA in $LiNbO_3$, noncollinear OPCPA in BBO and LBO, and noncollinear OPCPA in BBO pumped by fundamental, second- and third harmonics of an Yb:YAG thin-disk laser, respectively, are the methods of choice in terms of the amplification bandwidth and nonlinearity. These studies are exploited for experimental realization of a mJ-level, sub-two-cycle OPCPA system, centered at 1 μm (Chap. 5) and a design study of a multi-channel synthesizer (Chap. 6).

2.3 Tunable OPCPA

[1]Time resolved spectroscopy often requires tunable multi-cycle femtosecond pulses. OPCPA systems based on Yb:YAG lasers provide mJ-level, sub-50-fs pulses, tunable over a broad spectral range spanning from UV to MIR. In this section, the design and feasibility of such a system is simulated and discussed. The following simulation is performed for a Yb:YAG thin-disk regenerative amplifier yielding 30-mJ, 1.6-ps pulses at 10-kHz repetition rate, i.e. an average power level of 300 W (Teisset et al. 2013), as high average powers are often desirable in nonlinear spectroscopy to increase the signal-to-noise ratio.

The fundamental pulse along with its second and third harmonics, are used to pump and amplify a supercontinuum spanning from 450 nm to 3 μm. The seed is divided into three bands centered at carrier wavelengths of 550 nm, 1 and 2 μm, which are pumped with the third-harmonic, second-harmonic and fundamental pulses of the Yb:YAG laser, respectively. Each OPCPA arm consists of two amplifier stages, where the residual pump energy after the first stage is used to pump the second stage (see Sect. 5.1), using thin BBO, LBO and $LiNbO_3$ crystals, pumped at 343, 515 and 1030 nm, respectively.

[1]This part is adapted from (Fattahi et al 2014).

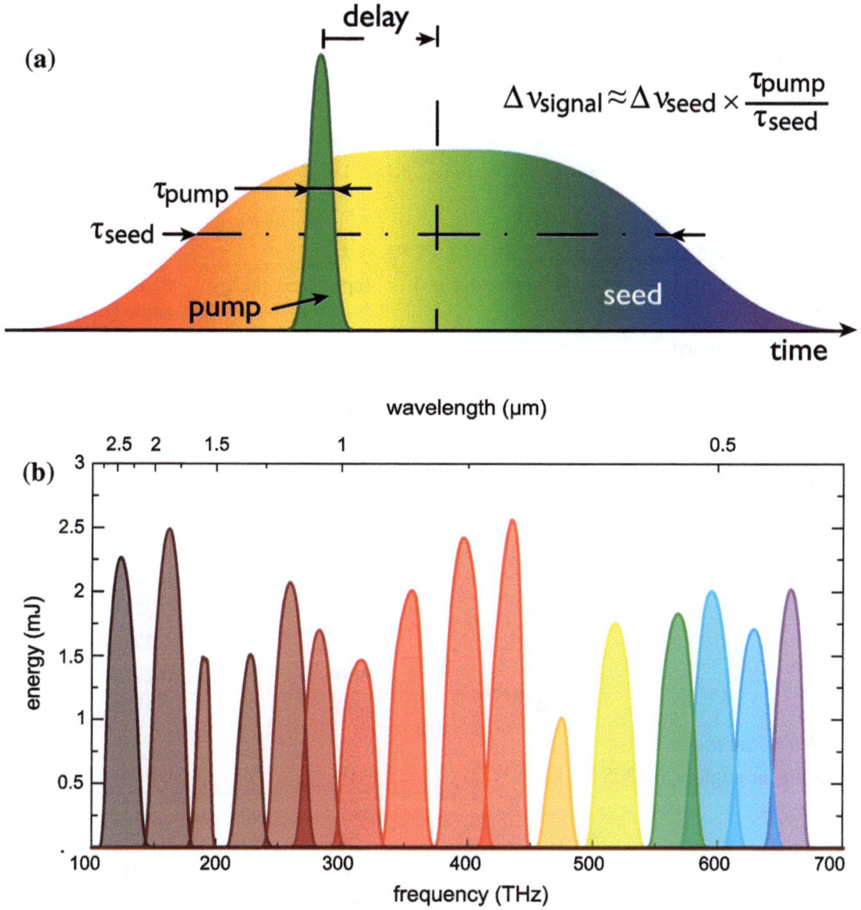

Fig. 2.21 Generation of widely tunable femtosecond pulses. **a** Scheme for generating spectrally and temporally tunable pulses at any wavelength. The seed pulses are stretched to a significantly longer duration than that of the pump pulse. Amplification in a simple OPCPA setup yields pulses that can spectrally be tuned by changing the temporal delay between pump and seed pulse. **b** Amplified spectra of a widely tunable two-stage OPCPA system obtained from simulations. Millijoule-level, sub-40-fs pulses, tunable from 445 to 2750 nm, can be generated by using different harmonics of a 10 kHz Yb:YAG regenerative amplifier. The shown spectra are normalized to their energy

Pulse-duration control and wavelength tuning of the amplified pulses is accomplished by temporally stretching the seed continua and controlling their delay with respect to the pump pulses. In fact, a pump temporal window of $\tau_{pump} \simeq 1$ ps slices out a fraction of the bandwidth, $\Delta\nu_{signal}$, of the stretched seed continuum, $\Delta\nu_{seed}$, which is inversely proportional to the duration τ_{seed} of the stretched seed: $\Delta\nu_{signal} \simeq \Delta\nu_{seed}(\tau_{pump}/\tau_{seed})$ (Fig. 2.21).

Thanks to the near-linear chirp carried by the stretched seed, the carrier frequency of the amplified signal can be tuned by varying the delay of the seed with respect to the pump pulse and by adjusting the phase-matching angle of the amplifier

Table 2.1 General parameters of the tunable OPCPA system

Pump harmonics	Pump energy (mJ)	Crystal	Tunable bandwidth (nm)
Fundamental	12	$LiNbO_3$	1500–2750
Second harmonic	10	LBO	670–1420
Third harmonic	8	BBO	445–660

Table 2.2 Detailed parameters of the tunable OPCPA system

	Wavelength (nm)	Input seed energy (nJ)	Phase-matching angle (rad)
Fundamental-pumped			
Noncollinear angle = 0	2450	300	0.7886
	1750	300	0.7893
	1550	300	0.8
Second harmonic-pumped			
Noncollinear angle = 0.0308	1350	150	0.26
	1150	105	0.2714
	1150	60	0.2714
	950	50	0.2714
	850	2	0.2714
	750	300	0.2714
	700	300	0.2714
Third harmonic-pumped			
Noncollinear angle = 0.1305	640	10	0.569
	575	50	0.5712
	525	50	0.5712
	500	300	0.5712
	475	100	0.5712
	450	300	0.5716

crystal. Figure 2.21 shows a series of amplified signal spectra from simulations in the three channels pumped with pulses of a duration $\tau_{pump} \simeq 1.7$ ps. The continuum is stretched to $\tau_{seed} \simeq 30$ ps.

For the numerical study of the tunable OPCPA system, a Gaussian beam profile for both pump and seed and a Gaussian temporal profile of the pump pulses are assumed. By using different harmonics of the regenerative amplifier and different crystals, mJ-energy, sub-40-fs pulses, tunable from 445–2750 nm could be generated. The length of the OPCPA crystals is kept constant for each chain. As the amplification gain is different for different frequencies, the gain saturation of the frequencies in one set of crystal lengths is ensured by shaping and scaling the amplitude of the input signal to the cascaded OPCPA stages. By utilizing a prism compressor, these pulses can be compressed close to their Fourier transform limits. The detailed parameters of the simulation are summarized in Tables 2.1 and 2.2.

Conclusion In this section a full design study of a simple, high-power widely tunable OPCPA source, pumped by a Yb:YAG regenerative amplifier is presented. Here, mJ, sub-40-fs pulses, tunable from 445 to 2750 nm at 10 kHz repetition rate can be generated. Tunability of the system is achieved by heavy temporal stretching of the seed pulses and adjustment of the relative delay between pump and seed pulses.

These synchronized femtosecond pulses with adjustable pulse durations and tunable carrier frequencies at such a variety of wavelengths and unprecedented average-power levels may open new prospects for sophisticated multi-dimensional spectroscopies and pump-probe schemes.

References

Augstov PA, Shvarts KK (1980) The temperature and light intensity dependence of photorefraction in LiNbO3. Appl Phys 21(2):191–194

Boyd RW (2003) Nonlinear optics

Bromage J, Rothhardt J, Hädrich S, Dorrer C, Jocher C, Demmler S, Limpert J, Tünnermann A, Zuegel JD (2011) Analysis and suppression of parasitic processes in noncollinear optical parametric amplifiers. Opt Exp 19(18):16797–16808

Cerullo G, De Silvestri S (2003) Ultrafast optical parametric amplifiers. Rev Sci Instrum 74(1):1

Deng Y, Schwarz A, Fattahi H, Ueffing M, Gu X, Ossiander M, Metzger T, Pervak V, Ishizuki H, Taira T, Kobayashi T, Marcus G, Krausz F, Kienberger R, Karpowicz N (2012) Carrier-envelope-phase-stable, 1.2 mJ, 1.5 cycle laser pulses at 2.1 μm. Opt Lett 37(23):4973–4975

Dmitriev VG, Gurzadyan GG, Nikogosyan DN (1999) Handbook of nonlinear optical crystals, p 414

Dubietis A, Tamošauskas G, Varanavičius A, Valiulis G (2000) Two-photon absorbing properties of ultraviolet phase-matchable crystals at 264 and 211 nm. App Opt 39(15):2437–2440

Fattahi H, Barros HG, Gorjan M, Nubbemeyer T, Alsaif B, Teisset CY, Schultze M, Prinz S, Haefner M, Ueffing M, Alismail A, Vámos L, Schwarz A, Pronin O, Brons J, Geng XT, Arisholm G, Ciappina M, Yakovlev VS, Kim D-E, Azzeer AM, Karpowicz N, Sutter D, Major Z, Metzger T, Krausz F (2014) Third-generation femtosecond technology. Optica 1(1):45–63

Furukawa Y, Kitamura K, Takekawa S, Miyamoto A, Terao M, Suda N (2000) Photorefraction in LiNbO3 as a function of [Li]/[Nb] and MgO concentrations. Appl Phys Lett 77(16):2494–2496

Gale GM, Cavallari M, Driscoll TJ, Hache F (1995) Sub-20-fs tunable pulses in the visible from an 82-MHz optical parametric oscillator. Opt Lett 20(14):1562–1564

Gale GM, Cavallari M, Hache F (1998) Femtosecond visible optical parametric oscillator. J Opt Soc Am B 15(2):702–714

Herrmann D, Tautz R, Tavella F, Krausz F, Veisz L (2010c) Investigation of two-beam-pumped noncollinear optical parametric chirped-pulse amplification for the generation of few-cycle light pulses. Opt Exp 18(5):4170–4183

Nakatani H, Bosenberg WR, Cheng LK, Tang CL (1988) Laser-induced damage in beta-barium metaborate. Appl Phys Lett 53(26):2587–2589

Pires H, Galimberti M, Figueira G (2014) Numerical evaluation of ultrabroadband parametric amplification in YCOB. J Opt Soc Am B 31(11):2608–2614

Riedle E, Beutter M, Lochbrunner S, Piel J, Schenkl S, Spörlein S, Zinth W (2000) Generation of 10 to 50 fs pulses tunable through all of the visible and the NIR. Appl Phys B 71(3):457–465

Rothhardt J, Demmler S, Hädrich S, Limpert J, Tünnermann A (2012) Octave-spanning OPCPA system delivering CEP-stable few-cycle pulses and 22 W of average power at 1 MHz repetition rate. Opt Exp 20(10):10870–10878

Schultze M, Binhammer T, Steinmann A, Palmer G, Emons M, Morgner U (2010) Few-cycle OPCPA system at 143 kHz with more than 1 microJ of pulse energy. Opt Exp 18(3):2836–2841
Schmidt BE, Thiré N, Boivin M, Laramée A, Poitras F, Lebrun G, Ozaki T, Ibrahim H, Légaré F (2014) Frequency domain optical parametric amplification. Nat Commun 5:3643–3648
Skrobol C, Ahmad I, Klingebiel S, Wandt C, Trushin SA, Major Z, Krausz F, Karsch S (2012) Broadband amplification by picosecond OPCPA in DKDP pumped at 515 nm. Opt Exp 20(4):4619–4629
Stuart B, Feit M, Herman S, Rubenchik A, Shore B, Perry M (1996) Nanosecond-to-femtosecond laser-induced breakdown in dielectrics. Phys Rev B 53(4):1749–1761
Thai A, Skrobol C, Bates PK, Arisholm G, Major Z, Krausz F, Karsch S, Biegert J (2010) Simulations of petawatt-class few-cycle optical-parametric chirped-pulse amplification, including nonlinear refractive index effects. Opt Lett 35(20):3471–3473
Teisset C, Schultze M, Bessing R, Haefner M, Prinz S, Sutter D, Metzger T (2013) 300 W picosecond thin-disk regenerative amplifier at 10 kHz repetition rate. In: Huber G, Moulton P (eds) Advanced solid-state lasers congress postdeadline, JTh5A.1. Washington, D.C., USA
Wnuk P, Stepanenko Y, Radzewicz C (2010) High gain broadband amplification of ultraviolet pulses in optical parametric chirped pulse amplifier. Opt Exp 18(8):7911–7916
Zhao B, Jiang Y, Sueda K, Miyanaga N, Kobayashi T (2008) Ultrabroadband noncollinear optical parametric amplification with LBO crystal. Opt Exp 16(23):18863–18868

Chapter 3
Broadband Seed Generation

A major requirement for developing the OPCPA systems discussed in Chap. 1, is the availability of seed pulses with a broadband spectrum, a well-behaved spectral phase, and preferably a stable carrier envelope phase (CEP).

Traditionally, OPCPA seeds were provided by using low-energy broadband spectrum of a Ti:sapphire oscillator Ishii et al. (2005). The seed pulses in this case have a good temporal contrast, well-behaved spectral phase and their carrier envelope phase offset can be actively stabilized Telle et al. (2014). By using the oscillator to seed the OPCPA chain and the OPCPA pump laser, pump and seed pulses are intrinsically synchronized Teisset et al. (2005). However due to the long optical beam path difference between seed and pump pulses, mechanical vibrations of optical components and temperature drifts additional timing jitter happens, which needs to be compensated by an active temporal synchronization system (Teisset 2009; Schwarz et al. 2012; Fattahi et al. 2012). Despite all the above mentioned advantages the pulse energy is limited to several tens of nJ. This can be resolved by using an additional OPCPA stage or generating the seed via supercontinuum after boosting the output of the oscillator in a regenerative amplifier. However, this results in more superfluorescence (Tavella et al. 2006) in the first approach and degraded temporal contrast in the second approach.

Recently demonstrated kerr-lens mode-locked Yb:YAG thin-disk oscillators (Pronin et al. 2011) are a promising alternative to Ti:sapphire as the front-end of OPCPA systems. Their relatively narrowband spectrum can be externally broadened in a cascaded nonlinear stage. Compression of the broadened spectrum results in 10 fs pulses with 260 nJ of energy at 40 MHz repetition rate (Seidel et al. 2014). The average power of these reproducible waveform-controlled few-cycle pulses exceeds the average power of Ti:sapphire oscillators by more than one order of magnitude. These externally CEP stabilized pulses (Pronin et al. 2013) combined with optical synchronization provides a suitable OPCPA seed at μJ-levels of energy.

The need for an active pump-signal temporal synchronization system can be eliminated by generating the broadband seed from a near-1-ps OPCPA pump pulses. In spite of several attempts to generate such a broadband spectrum via supercontinuum generation with sub-100-fs pump pulses in solids and waveguides (Seidel et al. 2014;

H. Fattahi, *Third-Generation Femtosecond Technology*,
Springer Theses, DOI 10.1007/978-3-319-20025-5_3

Bradler et al. 2009; Silva et al. 2012), the following question remains open: is there a straightforward approach to generate a broadband signal from a near-1-ps pulse? Another appealing question is: What is the best method for generation of a multi-octave, CEP-stable spectrum suitable for seeding a multi-channel OPCPA system?

In this chapter different methods for spectral broadening of sub-100-fs and near-1-ps pulses are discussed. In the beginning spectral broadening of near-1-ps pulses in large mode area (LMA) fibers is shown. Then sub-100-fs-pump spectral broadening in solids and photonic crystal fibers (PCF) are investigated. Cross-polarized wave generation (XPW) as a simple tool for pulse-shortening, which enhances the temporal contrast of the pulse simultaneously, are studied theoretically and experimentally. This chapter culminates with a full study of the difference frequency generation (DFG) process as the most promising route to the generation of a CEP-stable, multi-octave seed for OPCPA systems.

3.1 Spectral Broadening in Large Mode Area Fibers

Spectral broadening in a waveguide is preferred to spectral broadening in bulk solids due to the longer interaction length resulting in a higher total nonlinearity at a fixed intensity. In PCFs, pioneered by (P. St. J. Russell) in 1996 (Knight et al. 1997; Birks et al. 1997) dispersion and confinement of PCFs are determined by the periodicity of very tiny air holes, contrary to a normal step-index fiber, where the properties are designed by a spatially varying glass composition (Fig. 3.1). PCFs surpass the performance of step-index single-mode fibers, where their single-mode operation is limited to ~630–850 nm. Spectral broadening in the normal dispersion regime is dominated by SPM and Raman scattering, whereas in the anomalous regime, soliton fission Dudley and Coen (2006) is responsible for very broad continua. In the anomalous regime the spectrally broadened pulses have a complicated temporal structure which limits their compressibility. Therefore this regime is not an appealing approach to the generation of OPCPA seed pulses.

In the normal regime, LMA fibers have a higher nonlinearity compared to gas-filled HCFs and their higher damage threshold compared to single-mode fibers makes them suitable for the spectral broadening of high power, μJ-energy sources Südmeyer et al. (2003).

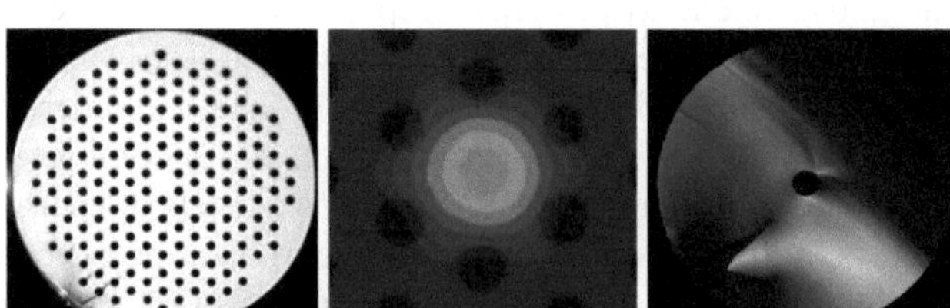

Fig. 3.1 Cross-section of an LMA fiber (*left*) and its beam profile in the far field (*middle*) (Newport Corporation 2014). Cross section of a HCF (*right*)

Experiment Figure 3.2 shows an experimental setup used for spectral broadening of 1 ps, 11 MHz, 5 μJ pulses of a Yb:YAG thin-disk oscillator (Fig. 3.3) Fattahi et al. (2012). First the output of the laser was focused into a 4 mm LBO crystal for SHG. The frequency-converted component of the beam was separated by a dichroic mirror and used later to pump an OPCPA stage, similar to Fattahi et al. (2012) and Teisset (2009).

The residual energy of the fundamental after the SHG stage, which still had a good beam quality (Fig. 3.3), coupled into a quartz LMA fiber with a zero dispersion wavelength at 1.3 μm. The fiber was placed in an aluminium V-groove on a 3-axis translation stage. The input energy to the fiber was adjusted by a tunable attenuator (half-wave plate and a thin film polarizer). A Faraday isolator was placed before the fiber to avoid any back-reflection from the entrance of the fiber to the laser.

Spectral broadening in fibers with different core diameters of 12, 15, and 20 μm were studied. Fig. 3.2a shows the broadened spectra in a 65 mm-long, 15 μm-core diameter LMA fiber for different input average powers. For average powers higher than 7 W and a fiber diameter smaller than 12 μm damage occurred. The spectrum broadened in a 12 μm-core-diameter LMA fiber supports 30 fs pulses. The pulses were collimated by using a 3 mm plano-concave lens and compression was performed in a chirped-mirror compressor with a GDD of $-500\,fs^2$ per reflection (Fig. 3.4b), producing a total GDD of $-6000\,fs^2$ over 12 reflections.

The total throughput of 57 % for the LMA fiber and the compressor left 4 W of average power for supercontinuum generation in bulk. The beam is focused into a 4 mm YAG crystal to generate a single filament with a spectrum reaching 500 nm (Table 3.1).

Figure 3.2c, d show the generated white light and its spectrum, in two different crystals: 4 mm-long YAG and 10 mm-long YVO_4. The average power of the white light was measured behind a high-reflectivity mirror at 1030 nm. The mirror was used to filter out the input spectrum and the scattered laser light. A white paper was used after the filter to diffuse the beam before coupling into the spectrometer. In both cases the efficiency of the supercontinuum generation did not exceed 2 %. The generated supercontinuum in YVO_4 has a narrower spectrum (Fig. 3.2d) and lower threshold due to the higher nonlinearity of YVO_4 crystal compared to the YAG crystal (Fig. 3.4a). Accordingly, multi-filamentation and damage occurred at lower pulse energies. The blue color, observable in the YVO_4 crystal in Fig. 3.2 is due to the simultaneous THG of the input pulse in the crystal.

The supercontinuum generation also was tried in sapphire. The spectrum and energy in this case was not stable as the peak power of the pulses was just above the critical peak power. Lower nonlinearity of sapphire compared to YAG and YVO_4 and hence its higher threshold demands a higher peak power to reach a stable regime of operation.

Conclusion Yb:YAG oscillator pulses with 7 W of average power and a spectral bandwidth of 1.4 nm, centered at 1030 nm are spectrally broadened in a cascaded broadening stage to a bandwidth supporting few-cycle pulses. Following broadening via SPM in a 12 mm-long LMA fiber and 12 reflections on a chirped-mirror

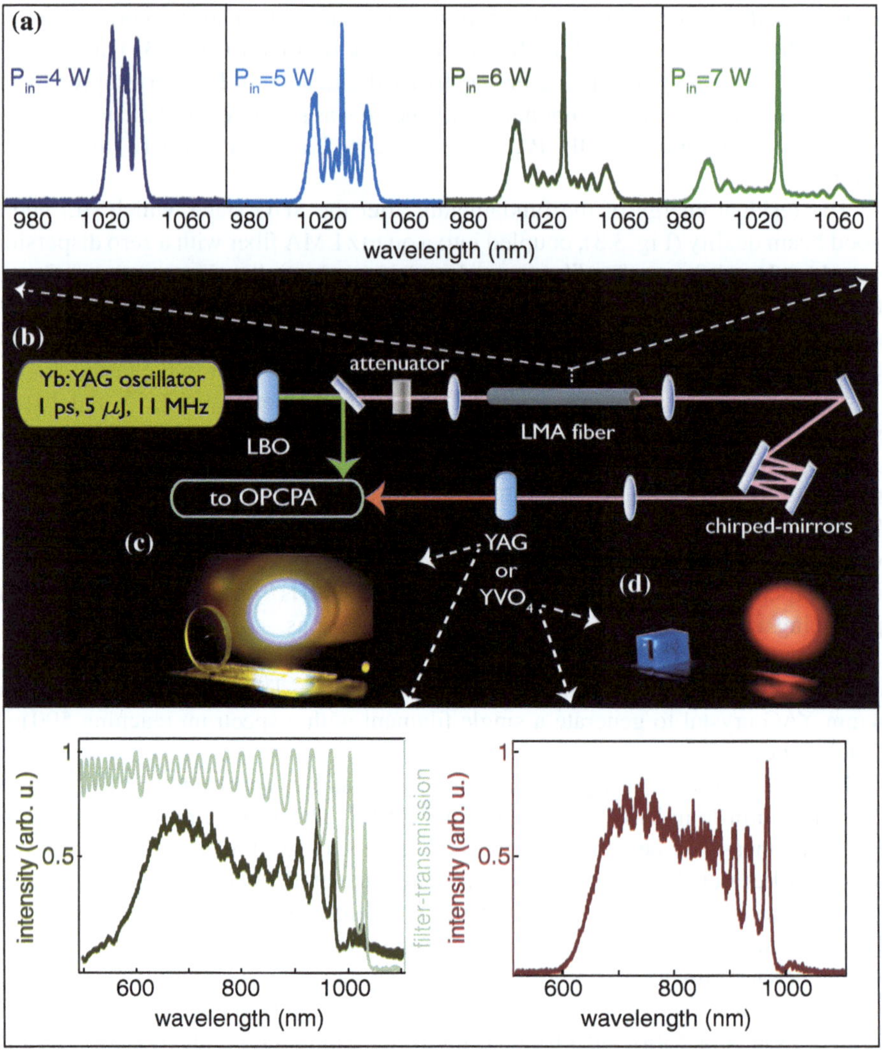

Fig. 3.2 **a** Spectral broadening in a 15 µm core-diameter-LMA fiber for different input average power normalized to one and plotted on a linear scale (P_{in} input average power). **b** Schematic layout of the experimental setup. The residual energy after frequency doubling of Yb:YAG oscillator pulses was spectrally broadened in an LMA fiber and later temporally compressed utilizing a chirped-mirror compressor. The compressed pulses are focused into a bulk material for supercontinuum generation which provides a coherent seed for optical parametric amplification. **c** Supercontinuum generation in a 4 mm-YAG crystal. The transmission curve of a filter used for spectrum and average power measurement is shown in light green. **d** Supercontinuum generation in a YVO_4 crystal. The generated blue light, visible in the picture of the crystal, is due to the simultaneous third harmonic generation of the input spectrum

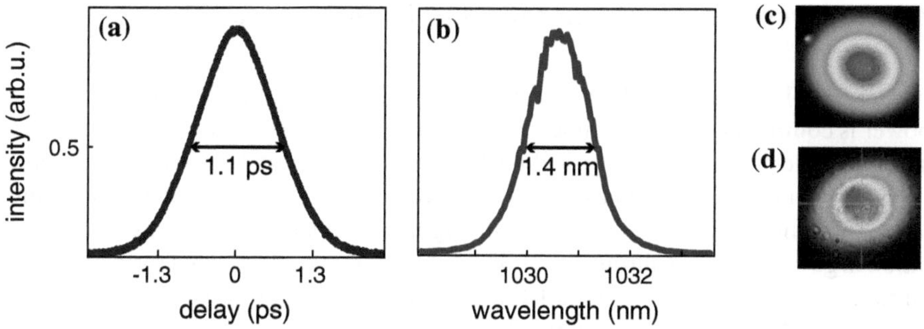

Fig. 3.3 Auto-correlation trace (**a**), spectrum (**b**) and the beam profile (**c**) of a SESAM mode-locked Yb:YAG thin-disk oscillator Fattahi et al. (2012). The beam profile of the residual of the fundamental after the SHG (**d**)

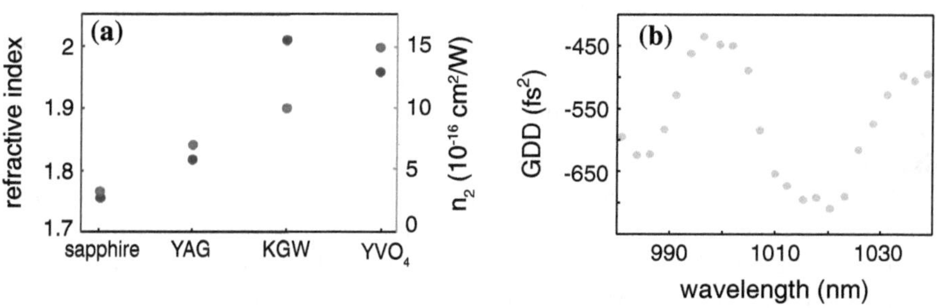

Fig. 3.4 **a** Refractive index and nonlinear refractive index of different bulk materials. **b** Measured GDD curve of the chirped mirrors used for the compression of the broadened spectrum generated in an LMA fiber

Table 3.1 Parameters of the supercontinuum generation in YAG and YVO_4 crystals

Crystal	Thickness (mm)	Input power (W)	Output power (mW)	Efficiency (%)
YAG	4	4	50	1.2
YVO_4	8	2.2	25	1.1

compressor, 30 fs pulses with 4 W of average power are obtained. Focusing these pulses into a 4-mm-thick YAG crystal results in a supercontinuum with 1 % of efficiency in the anti-Stokes wing. In a proof-of-principle experiment, the generated supercontinuum is amplified in a single OPCPA stage similar to Fattahi et al. (2012). As the signal and pump pulses are generated from the same laser, they are intrinsically synchronized. Therefore no additional temporal synchronization is required for temporal jitter compensation and amplification in the OPCPA stage. While this approach makes the setup simpler, it still suffers from the lack of carrier envelope phase stability of the generated supercontinuum and is not scalable in energy due to limitations of LMA fibers.

3.2 Supercontinuum Generation in Hollow Core Fiber

For large pulse energies, HCFs surpass the performance of PCFs since the optical power is confined to a large-diameter, gas-filled hole and therefore has a higher damage threshold intensity. In a gas-filled HCF, the process of broadening is initiated in the gas and the HCF acts as a dielectric waveguide which provides significantly longer nonlinear interaction length for the light in comparison to the case of filamentation in gas. The energy throughput is limited by damage and plasma formation at the entrance of the fiber as well as self-focusing in the waveguide. The latter limitation is overcome by differentially pumped fibers Robinson et al. (2006).

Broadening in a HCF takes place in two different regimes. For long pump pulses, purely SPM is observed, whereas for short pulses a more uniformly broadened spectrum indicates the influence of self-steepening Nisoli et al. (1997). Noble gases have high thresholds for ionization Nisoli et al. (1996) and are typically preferred. At very high gas pressures, energy loss due to ionization and defocusing at the fiber entrance reduces transmission and produces significant mode splitting, decreasing the bandwidth of the broadened pulses Wei et al. (2010).

Experiment 5 % of the negatively chirped output of a Ti:sapphire multipass amplifier Cavalieri et al. (2007) was passed through a pair of SF57 wedges for temporal compression down to 27 fs (Fig. 3.5). The compressed pulses with 60 nm spectral bandwidth at FWHM, centered at 790 nm, contained 30 μJ of energy. In an experimental setup shown in Fig. 3.6a, the beam was focused down to 84 μm and coupled into a 120 μm-core-diameter, 15 cm-long HCF placed on a V-groove in a gas cell, statically filled with 5 bar of krypton.

The gas cell was made of stainless steel with 1 mm Brewster-cut CaF_2 windows. The GDD of the input pulse, adjustable with SF57 wedges, and the pressure of krypton were optimised to obtain the maximum spectral broadening (Fig. 3.6b) and 50 % throughput in the HCF. The broadened spectrum at 4.3 bar was subsequently focused in a 4 mm YAG crystal to extend the spectrum to 1400 nm.

Conclusion The generated coherent broadband spectrum in a Kr-filled-HCF at 4.5 bar spans from 400 nm to 1000 nm and contains 15 μJ of energy. This spectrum is extended to 1400 nm by focusing the output of the HCF into a 4-mm-thick YAG crystal. These μJ-level broadband spectra are used to seed a 3-stage OPCPA system, discussed in Sects. 5.1 and 5.3.

3.3 Cross-Polarized Wave Generation

XPW Minkovski et al. (2004) is a degenerate four-wave mixing process which is the result of the interaction of intense, linearly polarized light with a nonlinear medium with an anisotropic third-order nonlinearity and (typically) an isotropic linear index.

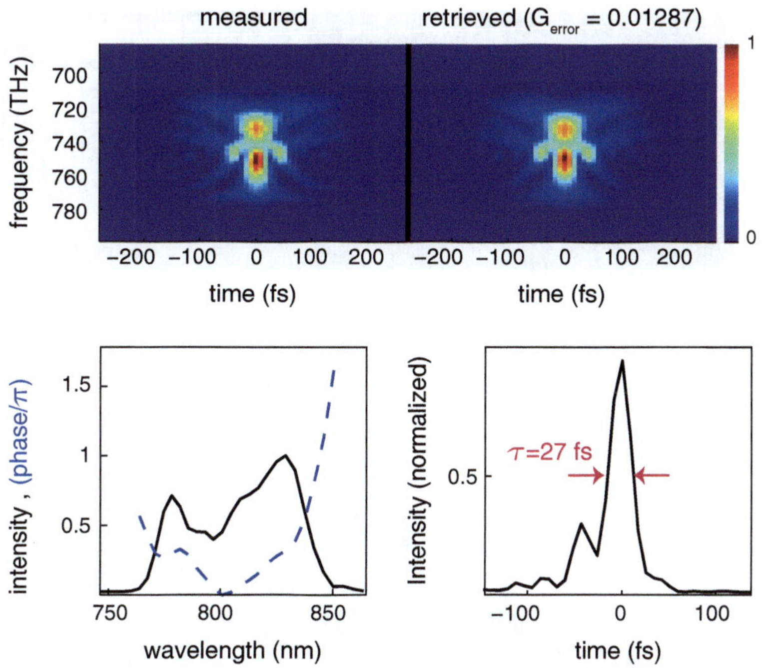

Fig. 3.5 Measured and retrieved SH-FROG (Second Harmonic-Frequency Resolved Optical Gating) traces (*top*) and the retrieved spectrum and temporal profile (*bottom*) of the pulses at the input of the HCF. The compressed pulse has the pulse duration of 27 fs at FWHM. The uncompressed pre-pulse pedestal, caused by the residual third order chirp, can be compensated by additional specially designed chirped mirrors

The XPW signal has an orthogonal polarization with respect to the pump and it is an automatically phase-matched process. XPW generation preserves the good spatial quality of the input beam and can improve the spatial profile of beams with distorted beam profiles. It also leads to significant shortening of the filtered pulse. The intensity of the generated XPW signal varies as the cubic power of the pump intensity. Suitable crystals for XPW generation are cubic crystals with Fm3m cubic symmetry (BaF_2, CaF_2, LiF). Among these crystals BaF_2 has a high $\chi^{(3)}$ anisotropy value which allows good conversion efficiency and its linear optical properties allow perfect group velocity matching along the propagation axis and a perfect spatial overlap of the cross-polarized waves. The crystal also has a wide transmission range from 0.2 to 11 μm (Cotel et al. 2006; Ricci et al. 2013).

Modulation in the frequency domain corresponds to satellite pulses in the time domain. In the XPW process these satellite pulses tend to be suppressed, as they have a lower intensity relative to the main pulse. Consequently, this pulse cleaning in the time domain translates to spectral smoothing since only the main peak of the pulse is converted. Therefore the temporal contrast of the generated pulse is greatly enhanced Jullien et al. (2005).

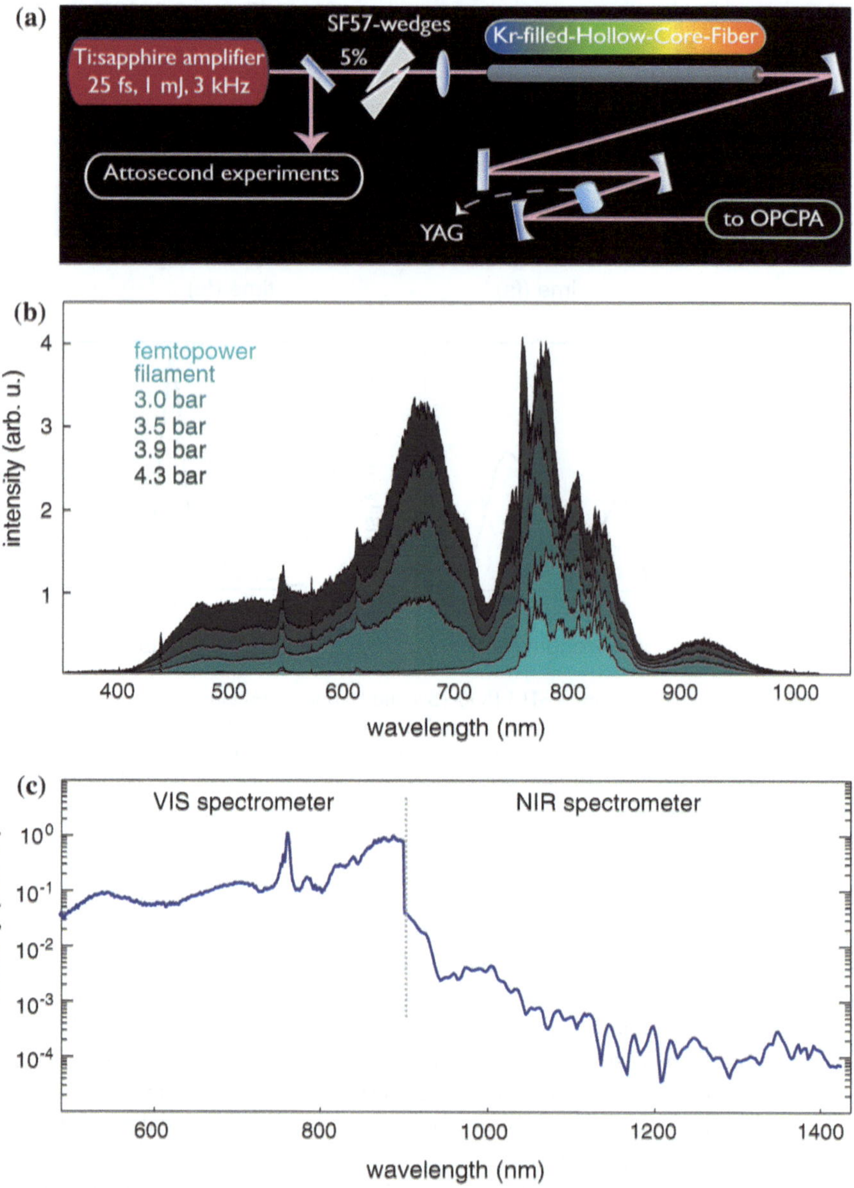

Fig. 3.6 **a** Layout of the seed generation scheme. A small portion of the energy of a Ti:sapphire multipass amplifier is focused into a 120 μm-core-diameter, Kr-filled HCF. The fine tuning of the group-delay-dispersion of the pulses was performed with a pair of SF57 wedges. **b** The spectrum of the Ti:sapphire multipass amplifier (Femtopower), the generated supercontinuum spectrum in Kr without HCF, and different spectra of the Kr-filled-HCF at different pressures. **c** In the second step, the broadband spectrum from the HCF filled with 4.3 bar of Kr, was focused into a 4 mm YAG crystal to extend the spectrum to 1400 nm.

XPW also acts as a pulse-shortening process. Here, unlike the other nonlinear phenomena, the generated XPW signal is always shorter than the pump pulse, as the signal is the result of the gating of the pump pulse in the time domain. For long pump pulses, the effect of dispersion of the crystal is negligible and the generated signal stays close to its Fourier transform limit.

In a linear crystal without birefringence, as long as changes in the nonlinear index of refraction can be neglected, the phase-matching condition is always fulfilled: $k_{\perp} = k_{\parallel} + k_{\parallel} - k_{\parallel}$.

Starting from Maxwell's equations and using the slowly varying envelope approximation, the following coupled wave equations for XPW are derived:

$$-i\frac{dA}{d\xi} = \gamma_1 AA^*A + \gamma_2 AA^*B + 2\gamma_2 AA^*B + 2\gamma_3 AB^*B + \gamma_3 BBA^* + \gamma_4 BB^*B, \quad (3.1)$$

$$-i\frac{dB}{d\xi} = \gamma_5 BB^*B + \gamma_4 BBA^* + 2\gamma_4 ABB^* + 2\gamma_3 ABA^* + \gamma_3 AAB^* + \gamma_2 AA^*A, \quad (3.2)$$

where A is the complex amplitude of the pump, B is the complex amplitude of the generated signal and ξ is the propagation direction. The coefficients in Eq. (3.1) and Eq. (3.2) describe different degenerate third-order processes that happen in the nonlinear medium simultaneously (Fig. 3.7) (Canova et al. 2008).

(a)

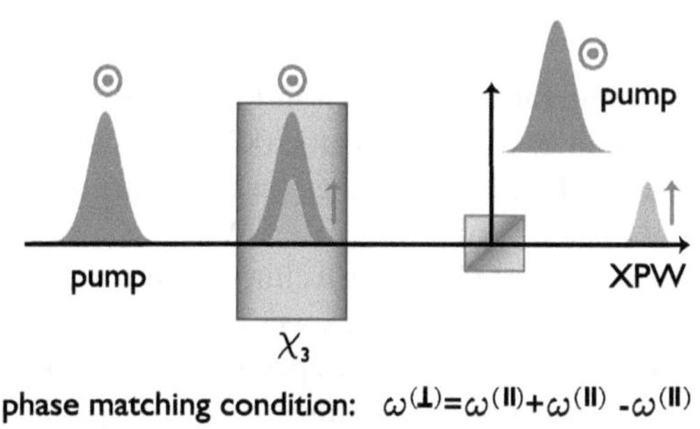

(b)

coefficient	γ_1, γ_5	γ_3	γ_2, γ_4
phenomena	self-phase modultion	cross-phase modulation	XPW

Fig. 3.7 **a** Schematic of crossed-polarized wave generation. Based on a degenerate four-wave mixing process in a nonlinear medium with an anisotropic third-order nonlinearity, a new wave with crossed-polarization with respect to the pump wave and significantly better temporal contrast is generated. **b** Role of each coefficient in the coupled wave equations describing the XPW process

To study the XPW process theoretically, the coupled wave equations are solved in a split-step method Taha and Ablowitz (1984). The pulse is propagated taking into account dispersion, diffraction and the third order nonlinearity, in the slowly evolving envelope approximation, in two dimensions (cylindrical coordinates with the assumption of radial symmetry). This is efficient enough to calculate not only the effects inside of one crystal, but also the spatio-temporal dynamics of the optimized two crystal systems discussed in Jullien et al. (2006). The pump pulse in the simulation is assumed to be Gaussian in space and time. 1 ps pulses centered at 1030 nm are propagated in a 2 mm, [011]-cut, BaF_2 at an intensity of 2 TW. It is assumed that $\chi^{(3)} = 1.59 \times 10^{-22}\,\mathrm{m}^2/\mathrm{V}^2$ Jullien et al. (2006) (Appendix B).

As can be seen in Fig. 3.8, in the first 0.5 mm length of the crystal, the pump pulse does not change its shape during propagation. Here the XPW signal varies as the cube of the input field. In this regime the pulse duration is reduced by a factor of $\sqrt{3}$. For a Fourier transform limited pulse, this translates to a spectral bandwidth increased by the same factor.

In a 1 mm-thick crystal, the efficiency of XPW reaches 12 %, with a smooth spectral broadening and pulse duration of 900 fs at FWHM. For longer crystal lengths, other third-order processes like SPM and cross phase modulation (XPM) take over XPW and the spectrum becomes modulated. At the same time, with the increase in efficiency of XPW, the pump pulse changes its shape in the time domain, and therefore the generated XPW spectrum becomes modulated. The introduced GDD from the SPM process and the simultaneous change of the shape of the pulse in the time domain makes the XPW pulse longer, while the Fourier transform limit pulse duration of XPW decreases with a sharper slope. This simulation shows that achieving high efficiencies in XPW comes at the expense of a modulated spectrum and therefore a decrease in the temporal contrast of the generated pulse Jullien et al. (2006).

Experiment In an experimental setup shown in Fig. 3.9, 95 % of the output of the Ti:sapphire multi-pass amplifier Cavalieri et al. (2007) was focused into a 1.1 m-long, 240 μm-core-diameter HCF, filled with neon gas at a pressure of 2 bar. The generated white light from the fiber had a total power of 1.5 W and a spectrum spanning from 400 to 1000 nm Cavalieri et al. (2007). Using a chirped-mirror compressor consisting of six mirrors and a pair of thin fused silica wedges, the pulses were compressed down to 4 fs with a total power of 1 W in the spectral range of 480–1000 nm.

The beam after the fiber was collimated using a $f = 1$ m spherical mirror and later by using a set of curved mirrors, the beam size is reduced to 400 μm. The telescope was misaligned to have a slightly divergent beam. The smallest beam diameter was limited by filamentation in air. To perform an intensity scan of the process, a 1 mm BaF_2 crystal is placed at different positions (P_0, P_1, P_2, P_3, P_4) along the divergent beam (Fig. 3.9). The pulses after the HCF are linearly p-polarized with no measurable energy in the s-polarization. Therefore no polarizer is used before the BaF_2 crystal. The generated XPW signal was separated from the pump by using a nanoparticle-linear-film polarizer (Thorlabs GmbH).

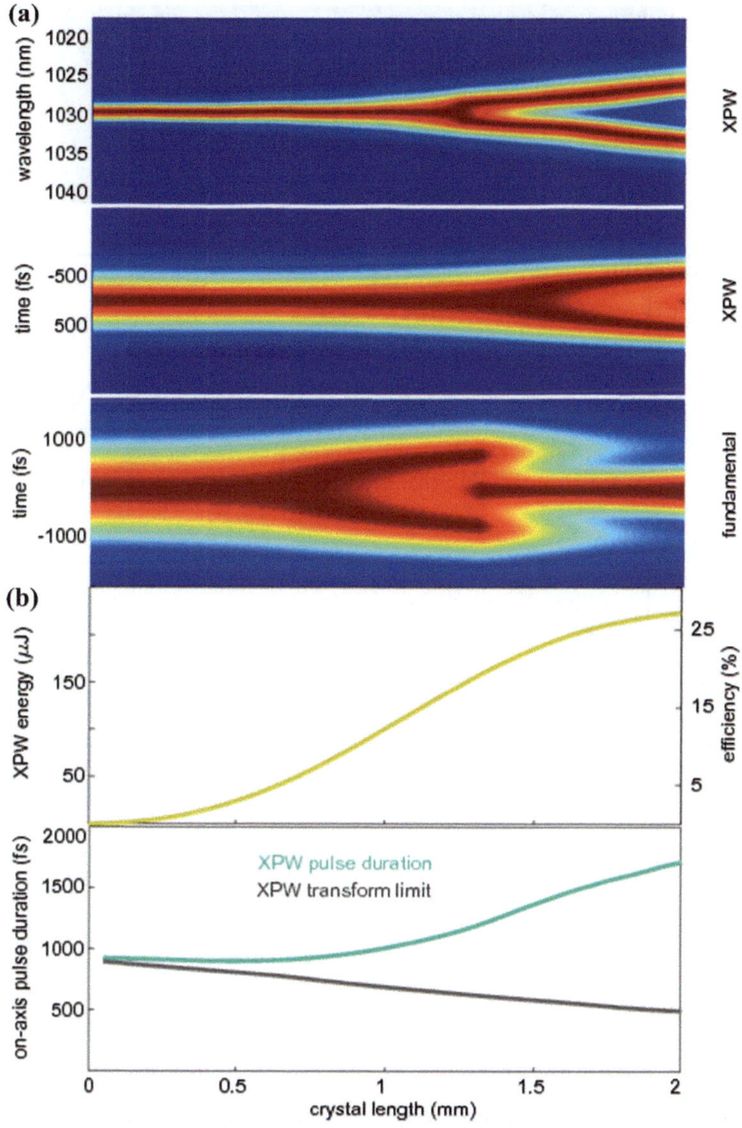

Fig. 3.8 **a** The generated XPW spectrum, its time domain profile and the temporal profile of the 1.6 ps, 1030 nm pump pulses propagating in a 2 mm BaF_2. **b** The signal reaches saturation at the end of the crystal, but after 1 mm of propagation, the spectrum becomes modulated due to SPM, XPM and backconversion of the energy to the pump pulse. Therefore the pulse becomes longer but the Fourier transform of the spectrum supports an even shorter pulse

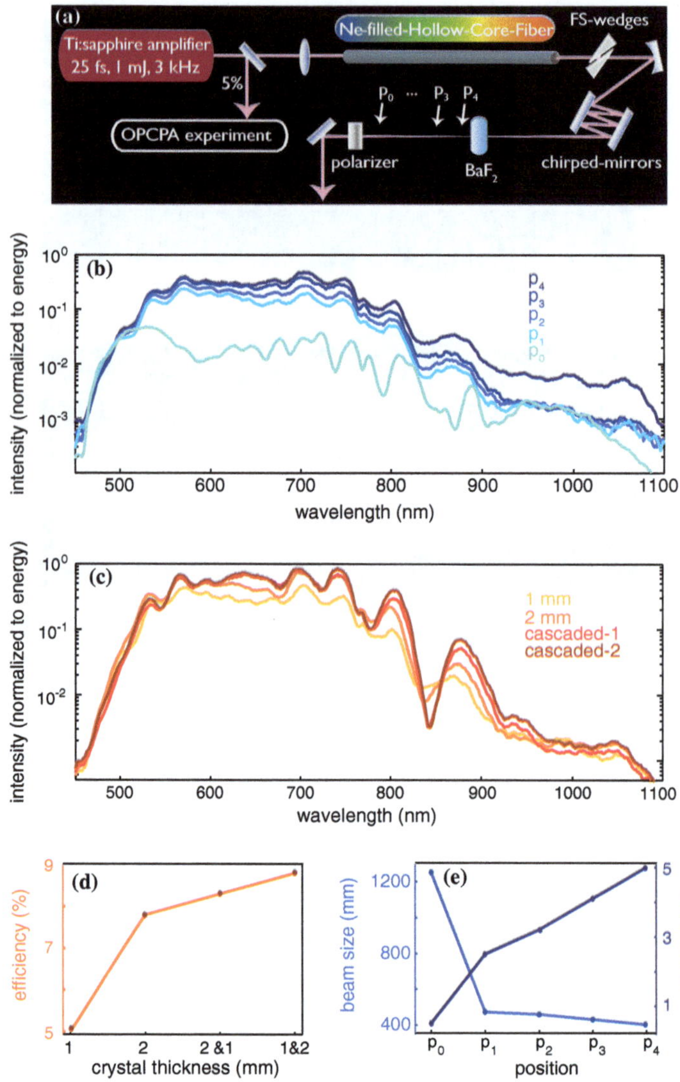

Fig. 3.9 **a** The experimental setup of XPW generation. **b** XPW spectra for different intensities. For the intensity scan the crystal was moved along a slightly divergent beam. P_i, where i: 0,…,4 indicates the position of the crystal on the pump beam (see the layout). **c** XPW spectra for different BaF_2 thicknesses and the cascaded case. "cascaded-1" shows the spectra of the crystal combination when the 2 mm crystal is placed before the 1 mm crystal and "cascaded-2" is the spectrum corresponding to the case where two crystals are switched. **d** and **e** The efficiency of the XPW signal for different intensities and crystal thicknesses

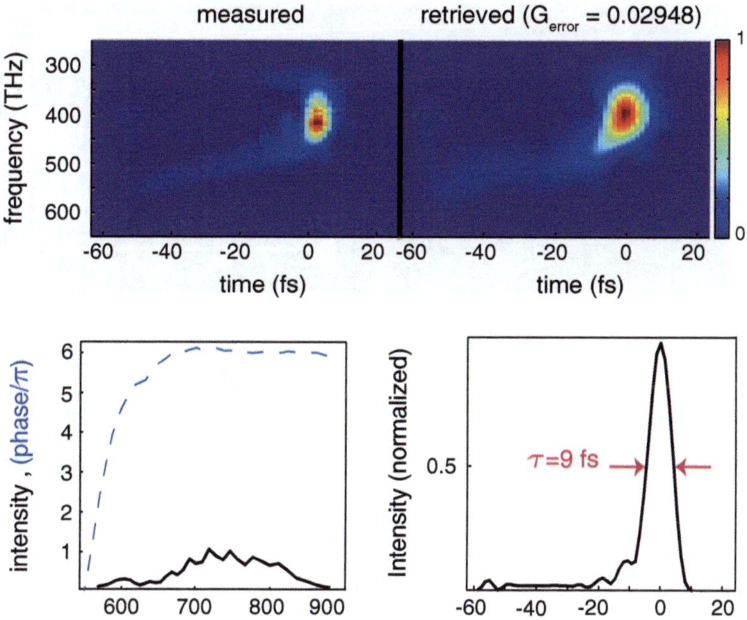

Fig. 3.10 Measured and retrieved TG-FROG traces (*top*) and the retrieved spectrum and temporal profile (*bottom*) of the measured pulses at the input of the XPW crystal. The compressed pulse has a duration of 9 fs at FWHM. The spectral component below 600 nm are not compressed which caused a long pedestal in the time domain

Generated XPW spectra and the corresponding efficiencies are shown in Fig. 3.9b, e. By increasing the intensity, the XPW spectrum became smoother and the modulations decreased. At position P_4, the red tail of the spectrum was enhanced which might be caused by other nonlinear processes.

In a second study, the crystal was fixed at position P_4 and BaF_2 with different thicknesses were used. Increasing the crystal thickness from 1 to 2 mm increased the efficiency of the process to 8 % (Fig. 3.9d).

Later a 1 mm crystal was placed after the 2 mm BaF_2 to increase the efficiency of XPW by taking advantage of the self focusing after the first crystal Cotel et al. (2006). The efficiency is slightly increased compared to the single crystal measurement. After switching the order of these two crystals, a yet higher efficiency was achieved as the pulse after propagation in the 1 mm crystal is less dispersed and the pump peak intensity in the second crystal remained higher. The spectrum of the latter combination is labeled with *cascaded-2* in Fig. 3.9c. The achieved extra efficiency in the cascaded crystal combination, is mostly dominated by extra photons in the red tail of the spectrum, while the blue tail of the spectrum was saturated.

In the third experiment, which was performed for low-energy pumped XPW (for the layout see Fig. 3.11), the spectrally broadened output of the HCF explained in Sect. 3.2 was compressed to 9 fs (Fig. 3.10) using 10 reflections in a chirped-mirror compressor. The temporally compressed pulses were characterized using a

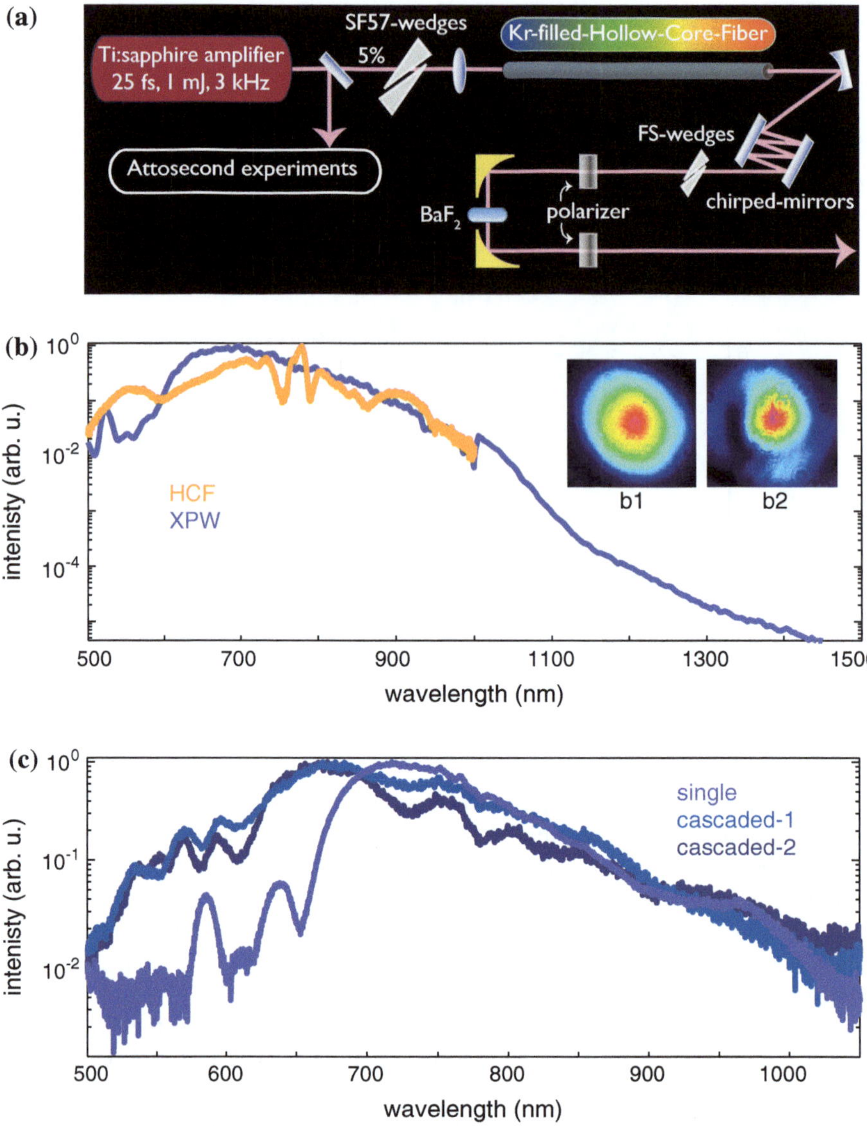

Fig. 3.11 **a** The experimental setup of XPW generation. **b** HCF spectrum (*orange*) and XPW spectrum (*purple*) generated in a 1 mm-thick BaF_2 crystal (both spectra are normalized to one). b1 demonstrates the spatial beam profile of the generated XPW signal and b2 shows the distortion of the XPW beam profile for very high peak intensity in the crystal. **c** Generated XPW spectra in the single crystal and cascaded crystals scheme. In the single crystal scheme, the blue tail of the spectrum has lower spectral intensity, as the input pulse is not compressed in this spectral range. The blue tail was enhanced for the case of the cascaded configuration (All spectra are normalized to one)

home-built transient grating frequency-resolved optical gating device (TG-FROG) utilizing a 500 μm fused silica plate as the nonlinear medium. The spectral components below 650 nm were not compressed as the design of the chirped-mirror compressor did not support this spectral range. A pair of thin fused silica wedges was used for fine tuning of the dispersion.

Unlike the previous study, in the high pressure gas-filled HCF, 6 % of the energy after the HCF was s-polarized, although the input pulse to the HCF was p-polarized. The generated s-polarized photons resulted from a crossed-phase Kerr effect. In a normal Kerr effect process, three photons with the same polarization are mixed and a fourth photon with the same polarization is generated: $P_{NL} \propto \chi^{(3)}_{pppp} E_p E_p E_p$. In the presence of the small number of s-polarized photons, crossed-phase Kerr effect, with the nonlinear polarization of $P_{NL} = \chi^{(3)}_{ppss} E_p E_p E_s$ results in mixing of two p-polarized photons with one s-polarized photon and generation of a new s-polarized photon. At the same time, two photons with s-polarization mix with a p-polarized photon and a new p-polarized photon is generated. In an infinitely long medium the process reaches equilibrium with equal number of s-polarized and p-polarized photons.

By adjusting the polarization of the input pulse to the HCF using a quarter-wave plate, the generated s-polarized photons were minimized to 0.3 % of the total average power. In order to enhance the polarization of the input beam before the XPW crystal, a nanoparticle-linear-film polarizer (Thorlabs GmbH) was placed after the chirped-mirror compressor. Then the beam was focused down to 200 μm at FWHM using a gold off-axis parabola. A BaF_2 crystal was placed on a translation stage directly after the focus to balance the self focusing in the crystal and the divergence of the beam in order to reach a uniform intensity over the crystal length. After the crystal, the beam is collimated using an identical off-axis gold parabola and sent through an identical nanoparticle-linear-film polarizer (Thorlabs GmbH) aligned for maximum transmission of the s-polarized light.

Close to the focus of the beam, the beam profile of the XPW signal was distorted (Fig. 3.11b1, b2). Using an uncoated, [011]-cut, 1 mm-thick BaF_2 crystal, 3.3 mW of average power (measured after the second polarizer) in an XPW signal was generated, corresponding to 12.5 % efficiency (Fig. 3.11b).

In a cascaded-crystal configuration similar to the first study, the generated XPW spectrum was enhanced in the blue tail of the spectrum and the efficiency was increased to 14 %. Besides the XPW procces, other third order phenomena like SPM and cross-phase modulation extended the red tail of the XPW spectrum to 1400 nm (Fig. 3.11b).

Conclusion In this section, few-cycle near-infrared pulses are used to generate a cross-polarized wave (XPW). The efficiency of the process is increased to 9 % by employing a cascaded-setup with 1 and 2 mm BaF_2 crystals. To reach a high peak intensity on the second crystal, two parameters are optimized: 1—temporal dispersion

Table 3.2 Comparison between different techniques to enhance the temporal contrast of ultrashort pulses

Technique	Cons
Plasma mirror	• Not suitable for high repetion rate lasers
Sagnac interferometer	• Hard to align • Complicated set up
Crossed-polarized wave generation	• Its contrast enhancement is limited to polarizer extinction ratio
Self diffraction	• The generated pulse suffers from angular dispersion

of the pump pulse entering the second stage is minimized by adjusting the length of the crystal in the first stage. 2—The distance between the two cascaded crystals is optimized in order to balance the resulting self-focusing of the pump pulse after the first stage and its divergence. It is observed that the generated XPW spectrum contains new frequencies in addition to the pump spectrum.

Beside XPW, other techniques like plasma mirrors Doumy et al. (2004), Sagnac interferometers Renault et al. (2005), and self diffraction Liu and Kobayashi (2010) can be used to enhance the temporal contrast of laser pulses (Table 3.2). Among these methods, XPW is chosen and implemented in an OPCPA setup described in Sect. 5.4 due to the possibility of generating coherent photons with new frequencies and enhancing the temporal contrast of the laser pulses simultaneously. Although XPW preserves CEP-phase stability of the pump pulses, a seed generation technique which produces intrinsically CEP-stable pulses combined with a good temporal contrast and a well-behaved spectral phase is desired. The next section presents a study of the DFG process, which meets both of these criteria, making it the ideal OPCPA seed.

3.4 Difference Frequency Generation

The combination of two properties in DFG makes it a popular source for ultrashort pulse generation: the ability to simultaneously phase-match a large fractional bandwidth and the passive CEP stability that can be obtained (Manzoni et al. 2004; Moses et al. 2012; Homann et al. 2012)[1]. However, the conversion efficiency into the long wavelengths are usually low, especially in broadband DFG, where the crystal thickness has to be kept short.

[1]This section is adapted from Fattahi et al. (2013).

In this section different types of phase-matching for generating a broadband signal via DFG are experimentally and theoretically discussed. It is shown that a few-cycle VIS-NIR pulse obtained from spectral broadening of a commercial Ti:sapphire laser can be used to efficiently obtain broadband pulses in the MIR by implementing a simple method. Extending the CEP-stable broadband MIR pulses to VIS is the topic of Sect. 6.1.

Theory There are three ways of phase-matching for single-pulse DFG in BBO crystals. type I ($e - o \rightarrow o$) and type II ($e - e \rightarrow o$), ($e - o \rightarrow e$). These three cases and their corresponding phase-matching angles are depicted in Fig. 3.12.

A one-dimensional propagation code based on SVEA Brabec and Krausz (1997) which includes only second-order nonlinearities, is used for theoretical study of the DFG process. Figure 3.12e compares the generated DFG signal from a 3.5 fs Gaussian input pulse in a 500 μm-thick BBO. In ($e - e \rightarrow o$) type II and ($e - o \rightarrow o$) type I phase-matching, the generated spectrum has more than four times the bandwidth of ($e - o \rightarrow e$) type II phase-matching. Therefore the following study is limited to the two broadband phase-matching configurations.

In type I DFG, the underlying process is very similar to optical parametric amplification. The blue spectral components in the pulse on the fast axis transfer energy to the red components on the slow axis. During this process a phase-stable idler wave is generated and the phase-matching condition is given by:

$$k_p - k_s = k_i \tag{3.3}$$

where k_p, k_s and k_i are the wave vectors of the blue, red, and the idler wave respectively. As the process is similar to OPA, increasing the number of photons on the fast axis, or pump wave, results in a higher efficiency. Figures 3.13 and 3.14b show the efficiency curves of the DFG process and its corresponding spectrum in a type I BBO crystal for three different polarization orientations of the pump beam relative to the extra-ordinary axis of the crystal.

The highest achieved efficiency is at $\alpha = 30°$ where a larger number of photons of the incoming beam are projected onto the extraordinary axis of the crystal. This suggests that the process should become even more efficient by changing the spectral distribution of the input pulse on different axes of the crystal. That is, by sending the blue photons to the extraordinary axis of the crystal and red photons to the ordinary axis. In the presence of a quartz waveplate, the blue spectral components of the pulse obtain more phase shift than its red spectral components and they become separated. Consequently, by implementing a thin waveplate before the BBO crystal the fraction of photons with shorter wavelength on the fast axis can be increased which leads to the rise in the efficiency of the DFG process.

Type II DFG is very similar to optical rectification in the terahertz spectral domain Bakker et al. (1998): the nonlinear polarization takes the form of the intensity profile of the pulse. By multiplying the phase-matching condition, given in Eq. (3.3) by c:

	type I (e-o → o)	type II (e-e → o)	type II (e-o → e)
Θ	21.4°	40°	25°
Φ	90°	0°	0°

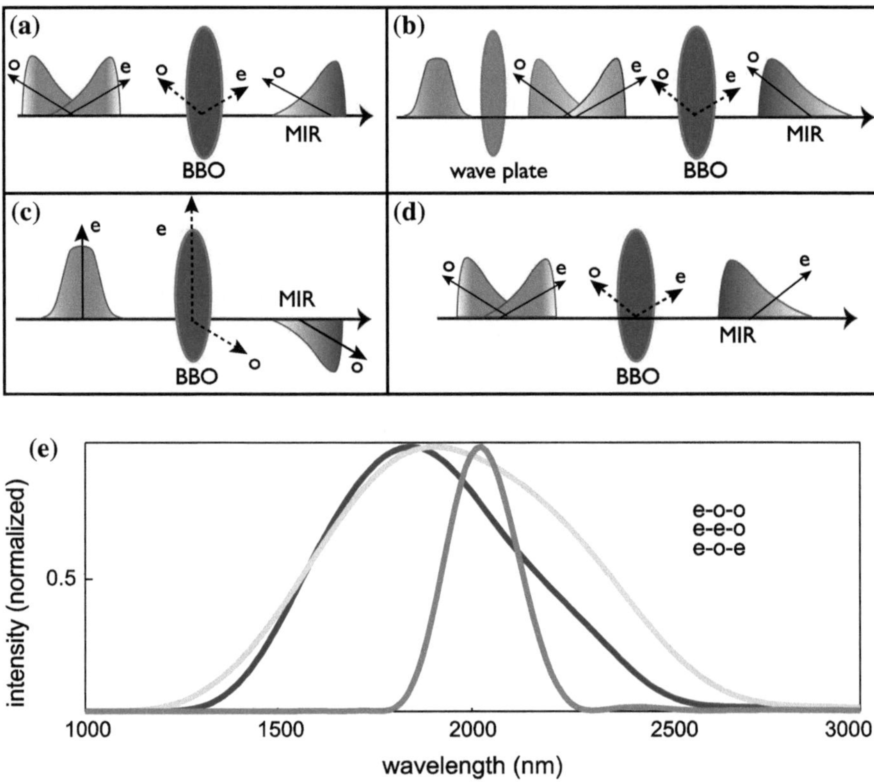

Fig. 3.12 Diagram of: **a** type I phase-matching with identical pulses sent to the fast and slow axis of a BBO crystal, **b** type I phase-matching using a wave plate to optimize the short-wavelength spectral components on the fast axis, **c** type II phase-matching with a pulse propagating along the fast axis of the crystal, and **d** type II phase-matching with a pulse propagating along the slow and fast axis of the crystal. **e** Simulated DFG spectra for the three phase-matching possibilities. The phase-matching angle for different types of phase-matching in a BBO crystal is summarized in the table

$$\omega_p n_p - \omega_s n_s = \omega_i n_i, \tag{3.4}$$

for broadband pump and signal with the angular frequency of ω_o and the spectral width of Δ:

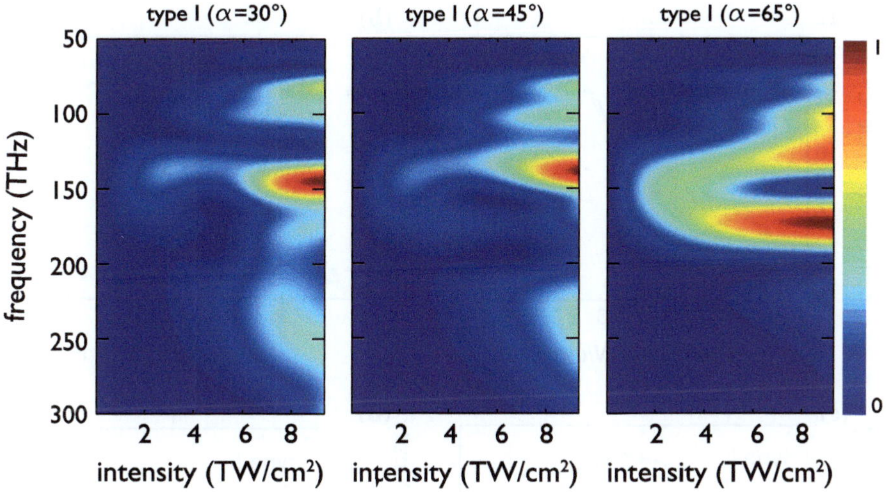

Fig. 3.13 Simulated spectra for three different orientations of the crystal relative to the input beam polarization. α shows the angle between the polarization of the incoming beam and the optical axis of the crystal

$$\omega_p = \omega_o + \Delta, \tag{3.5}$$
$$\omega_s = \omega_o - \Delta, \tag{3.6}$$
$$\omega_i = \omega_p - \omega_s = 2\Delta. \tag{3.7}$$

The refractive indices of the signal and pump can be expressed to first order in their series expansions

$$n_p = n_o + \Delta \left.\frac{dn_o}{d\omega}\right|_{\omega_o}. \tag{3.8}$$
$$n_s = n_o - \Delta \left.\frac{dn_o}{d\omega}\right|_{\omega_o}. \tag{3.9}$$

By substituting Eqs. (3.5) and (3.8) in Eq. (3.4):

$$\omega_i n_o + \omega_i \omega_o \left.\frac{dn_o}{d\omega}\right|_{\omega_o} = \omega_i n_i, \tag{3.10}$$

and by dividing the Eq. (3.10) by ω_i:

$$n_o + \omega_o \left.\frac{dn_o}{d\omega}\right|_{\omega_o} = n_i, \tag{3.11}$$

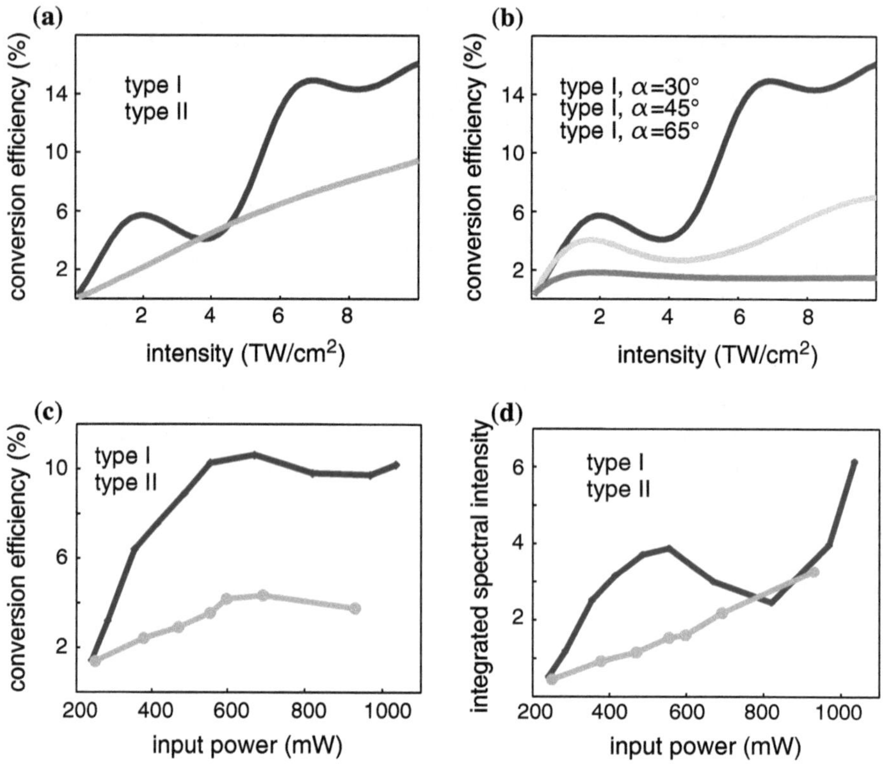

Fig. 3.14 Simulated conversion efficiency of light for **a** type I and II phase-matching, **b** type I phase-matching with different ratios of the beam on each optical axis of the 500 μm-thick BBO crystal. α shows the angle between the polarization of the incoming beam and the extra-ordinary axis of the crystal. **c** Measured scaling of DFG conversion efficiency versus pump power. **d** Measured integrated spectral intensity as a function of pump power. The pump power was adjusted with an iris and the spectrum was measured for a beam much larger than the spectrometer entrance slit, indicating the on-axis scaling of the MIR pulse

which is the expression for group index. By neglecting higher order terms, we obtain:

$$n_g^{pump} = n_0^{DFG}. \tag{3.12}$$

This phase-matching expression shows that the group velocity of the pump pulse should be equal to the phase velocity of the emitted wave. Here the dipole responsible for nonresonant DFG is simply the intensity envelope of the input laser pulse ($P^{NL}(t) \propto E^*(t)E(t) \propto I(t)$), so matching its velocity to the wavefronts of the DFG pulse results in phase-matching.

The expression also shows intuitively why the ($e - e \rightarrow o$) type II process is more broadband than the ($e - o \rightarrow e$) one, namely that the dispersion along the fast axis is greater (more negative) in the infrared than on the slow axis, reducing the range of frequencies where phase-matching is approximately satisfied.

The simulated conversion efficiencies in 500 µm-thick BBO crystals cut for type I ($\theta = 21.4°$, $\phi = 90°$, light polarization 30° from fast axis) and type II ($\theta = 40°$, $\phi = 0°$) are shown in Fig. 3.14. Here, at low pump intensities for unsaturated DFG, the efficiency is linearly proportional to the pump intensity. Increasing the pump intensity leads to higher conversion efficiency. In type I DFG, at high pump intensity, the linearity is broken and the DFG process saturates as the energy transfer from the generated MIR pulse to the pump pulse through sum frequency generation is faster than energy transfer from the pump to the DFG signal. Due to extremely short pulses in a single-pulse DFG studied here, the effect of group velocity matching results in rather different behavior of the two phase-matching types at high conversion efficiencies.

In a type I process the group velocity walk-off between the pulses is much less than in type II, which leads to a clear oscillatory behavior in the efficiency curve. Let us consider the case where the MIR wavelength is 2000 nm, with effective signal and pump wavelength of 800 and 580 nm. The group indices Kato (1986) of DFG, signal and pump are 1.6750, 1.6845, and 1.6932, respectively. Therefore the propagation distance in the crystal for a group delay shift of 10 fs between DFG and signal is 315.5 µm, and for the same shift between signal and pump is 344.6 µm. In the case of type II phase-matching, the DFG, signal and pump group indices are 1.6750, 1.6313, 1.6549, respectively, leading to much shorter distances to separate the DFG from the signal (68.6 µm) and the signal from the pump (127.0 µm). These relatively short length scales reduce the efficiency of the type II process, but at the same time ensure that backconversion through sum-frequency generation is unlikely at realistic intensities.

The effect of saturation on the DFG spectrum is illustrated in Fig. 3.15. As the intensity is increased, the peak wavelength of the DFG spectra obtained from type I phase-matching exhibits a shift first to longer and then to shorter wavelengths once the local minimum in conversion is reached. The spectral changes in type II phase-matching are more stable and the long-wavelength components are gradually added to the spectrum at higher intensities. This is caused by the overall redshift of the input pulse as energy is transferred from the short-wavelength side of the input pulse spectrum to the long-wavelength side, shifting the phase-matching to suit lower frequencies, as shown in Fig. 3.16.

The measurements were performed for both types of phase-matching and the energy of the generated DFG and its corresponding spectrum were measured separately for different input pump energies, scaled by adjusting a pinhole placed before the BBO crystals. The on-axis energy of the DFG signal was calculated by integrating over each DFG spectra, measured for different pump energies (Fig. 3.14d). The maximum measured DFG efficiency for type I BBO crystal was 11 % and for type II, it was 4.4 %. The obtained spectra for each phase-matching scheme and for different pump intensities are shown in Fig. 3.15.

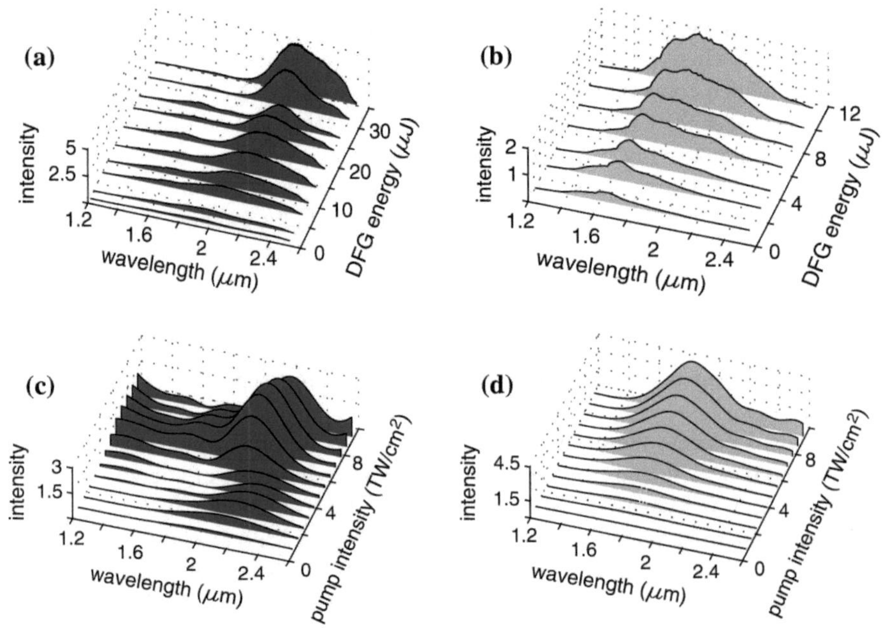

Fig. 3.15 Measured and simulated DFG spectra in a 500 μm-thick BBO crystal for type I (**a**, **c**) and type II (**b**, **d**) phase-matching at different intensities

Experiment 4 fs pulses with 1 W of average power Cavalieri et al. (2007) were loosely focused down to 0.4 mm at FWHM by using a spherical mirror with 1 m focal length. Tighter focusing of the beam resulted in a filament in air and was therefore avoided in the experiment. The 500 μm-thick, antireflection-coated BBO crystals were placed in the laser focus. A spherical mirror was used for collimating the beam after the BBO and several gold coated mirrors were used to either send the beam to a spectrometer or to a power-meter. In order to block the residual short-wavelength light, a 2 mm-thick antireflection-coated silicon wafer was placed in the beam to separate the generated DFG from the fundamental beam (Fig. 3.17).

To optimize the efficiency in type I phase-matching, a 42 μm-thick quartz plate was placed far from the focus and at Brewster's angle, such that the short-wavelength components were predominantly on the fast axis of the BBO. By adjusting the quartz plate, the DFG efficiency was increased to 12.4 %. The measured spectrum corresponding to the highest DFG pulse energy(41 μJ), is shown in Fig. 3.17b, alongside the spectrum of the pump pulses.

In the generation of broadband single-pulse DFG signals, a short interaction length in the material is required to avoid separation between different frequencies in the time domain, caused by dispersion in the crystal. The short pulses used in this experiment are important for reaching high conversion efficiencies as crystals with short thicknesses, could be exposed to higher intensities for reaching higher conversion efficiency. This would not be possible with longer pulses Lenzner et al. (1998).

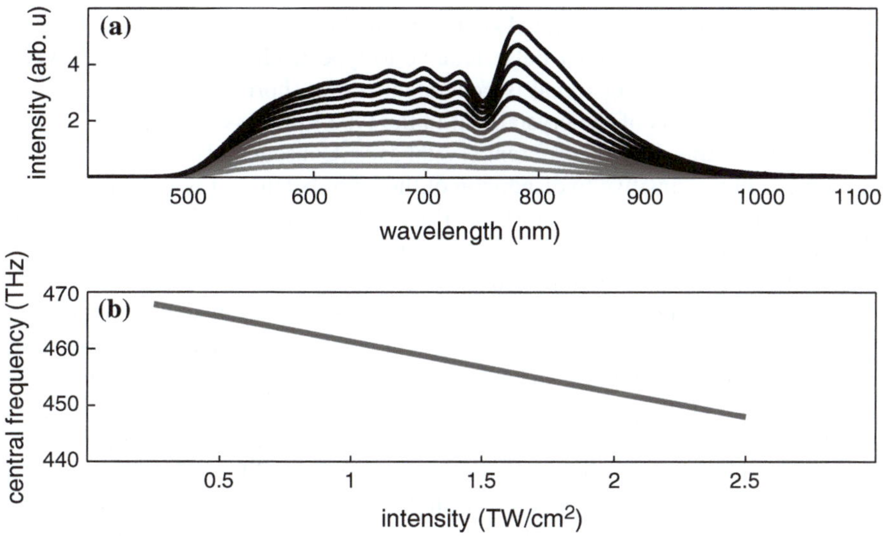

Fig. 3.16 Simulated redshift of the pump pulse with increasing intensity. **a** Output spectra of the pump pulse emerging from the e axis of the crystal with increasing intensity (from 0.3 to $2.5\,\mathrm{TW/cm^2}$). **b** The first moment frequency of the spectra shown in panel (**a**) versus intensity

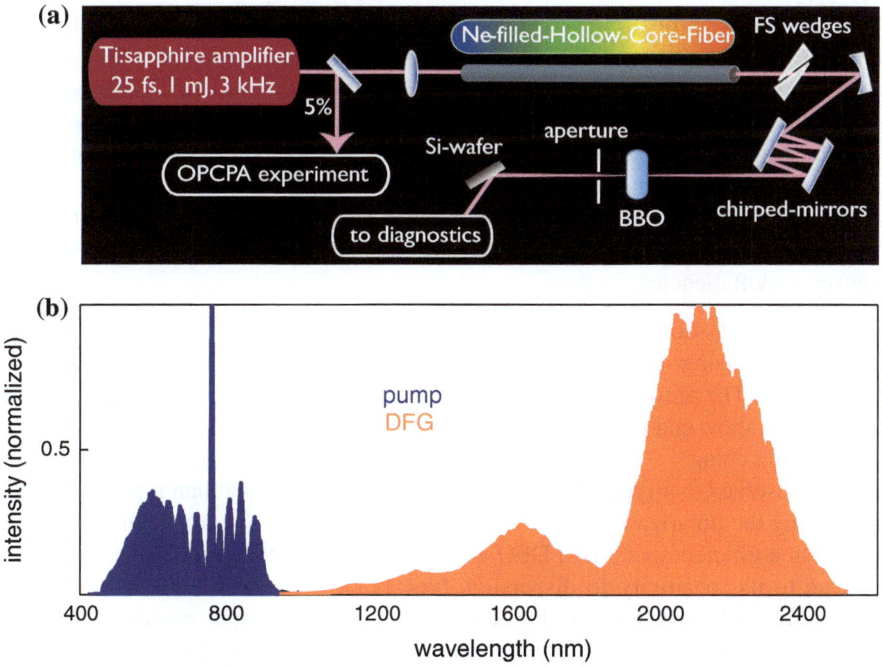

Fig. 3.17 **a** Layout of measurement set-up. **b** Pump spectrum after chirped-mirror compressor and the DFG spectrum with highest efficiency

In the experiment, the generated signal in type I phase-matching was less dependent on the chirp of the input pulse relative to type II phase-matching. In type II phase-matching, the DFG signal is generated from a short pulse which travels on the extraordinary axis of the crystal, but in the type I process the input pulse splits into two beams traveling on different crystal axes. Consequently, the type II phase-matching is more sensitive to dispersion of the input beam as it is very important to have all the frequencies at the same time. But in type I phase-matching the interaction happens between two relatively long pulses traveling on different axis of the crystal and it is less sensitive to the input pulse chirp.

The output power of the type II process had higher stability than the type I process. This behavior can be explained by looking at the conversion efficiency of both cases. In type II phase-matching, the behavior of the efficiency is linear and therefore the energy stability of the generated signal is proportional to the energy stability of the pump pulse. In contrast, for type I phase-matching several local saturation points appear for different intensities. In this case the most stable regime for operation is slightly above saturation, where for slightly different input pulse energies, the same output energy is obtained. However, the shape of the generated spectrum is not robust against intensity fluctuations. The backconversion of the DFG signal to the input signal changes the shape of the input in the time domain and its corresponding spectrum. This can cause spectral instabilities in the generated signal as shown in Fig. 3.15.

The energy stability of type II DFG would make it more attractive, despite its lower conversion efficiency. Also it is an advantage that the emerging white light and MIR pulses are orthogonally polarized: the residual pump light can be non-dispersively suppressed, via a Brewster's angle reflection from a silicon wafer, which we found to reduce the residual white light contamination level to a few percent, dominated mostly by residual higher-order modes from hollow-core fiber.

The damage threshold of the BBO was not reached in the measurement so performing the experiments in vacuum with a smaller focus could lead to yet higher conversion efficiencies.

Conclusion As shown in this chapter, difference frequency generation in a type I BBO crystal using 4.5 fs pulses centered at 800 nm yields conversion efficiencies above 10 %. The presented experimental observations and corresponding numerical simulations show qualitatively different intensity-scaling behavior for type I and type II phase-matching.

It is observed that type I DFG has higher overall efficiency than type II DFG. By modifying the polarization state of the pulse using a 42 μm zero-order wave plate, conversion efficiency of type I DFG was improved to 12 %.

Due to its sensitivity to dispersion, type II DFG has a lower conversion efficiency but enables better stability of the spectrum versus input intensity. The DFG signal was separated from the pump pulses entirely by using a polarizer, as it had a different polarization. These two characteristics of type II DFG, stability and its simple separation, makes it a suitable choice for producing broadband, CEP stable, MIR pulses with enhanced temporal contrast. By focusing the generated pulses in a gas-filled hollow-core-fiber Schwarz (2014) the spectrum extends to the VIS range. The remarkable role of such a multi-octave, CEP-stable spectrum to seed an OPCPA system (discussed in Chap. 6) leaves DFG an insuperable technique for generation of broadband, coherent spectrum with the carrier phase locked to the intensity envelope.

References

Ishii N, Turi L, Yakovlev VS, Fuji T, Krausz F, Baltuska A, Butkus R, Veitas G, Smilgevicius V, Danielius R, Piskarskas A (2005) Multimillijoule chirped parametric amplification of few-cycle pulses. Opt Lett 30(5):567–569

Telle HR, Steinmeyer G, Dunlop AE, Stenger J, Sutter DH, Keller U (2014) Carrier-envelope offset phase control: a novel concept for absolute optical frequency measurement and ultrashort pulse generation. Appl Phys B 69(4):327–332

Teisset C, Ishii N, Fuji T, Metzger T, Kohler S, Baltuska A, Krausz F, Zheltikov AM (2005) All-optical pump-seed synchronization for few-cycle OPCPA. In: (CLEO) Conference on Lasers and Electro-Optics, 2005, pp. 25–27. vol 1. IEEE

Tessiet C (2009) Few-cycle high-repetition-rate optical parametric amplifiers and their synchronisation schemes. PhD thesis

Schwarz A, Ueffing M, Deng Y, Xu G, Fattahi H, Metzger T, Ossiander M, Krausz F, Kienberger R (2012) Active stabilization for optically synchronized optical parametric chirped pulse amplification. Opt Express 20(5):5557–5565

Fattahi H, Teisset CY, Pronin O, Sugita A, Graf R, Pervak V, Xun G, Metzger T, Major Z, Krausz F, Apolonski A (2012) Pump-seed synchronization for MHz repetition rate, high-power optical parametric chirped pulse amplification. Opt Express 20(9):9833–9840

Tavella F, Marcinkevičius A, Krausz F (2006) Investigation of the superfluorescence and signal amplification in an ultrabroadband multiterawatt optical parametric chirped pulse amplifier system. New J Phys 8(10):2191–21911

Pronin O, Brons J, Grasse C, Pervak V, Boehm G, Amann M-C, Kalashnikov VL, Apolonski A, Krausz F (2011) High-power 200 fs Kerr-lens mode-locked Yb:YAG thin-disk oscillator. Opt lett 36(24):4746–4748

Seidel M, Brons J, Fedulova E, Pervak V, Apolonski A, Pronin O, Krausz F (2014) High-power few-cycle pulse generation by spectral broadening in bulk material. In: CLEO: 2014 Postdeadline Paper Digest. OSA, Washington, D.C., STh5C.9

Pronin O, Seidel M, Brons J, Lücking F, Pervak V, Apolonski A, Udem T, Krausz F (2013) Carrier-envelope phase stabilized thin-disk oscillator. In: Huber G, Moulton P (ed) Advanced solid-state lasers congress. OSA, Washington, D.C. AF3A.5

Bradler M, Baum P, Riedle E (2009) Femtosecond continuum generation in bulk laser host materials with sub-μJ pump pulses. Appl Phys B 97(3):561–574

Silva F, Austin DR, Thai A, Baudisch M, Hemmer M, Faccio D, Couairon A, Biegert J (2012) Multi-octave supercontinuum generation from mid-infrared filamentation in a bulk crystal. Nat commun 3(May):8071–8075

Knight JC, Birks TA, Russell PS, Atkin DM (1997) All-silica single-mode optical fiber with photonic crystal cladding: errata. Opt Lett 22(7):484–485

Birks TA, Knight JC, Russell PS (1997) Endlessly single-mode photonic crystal fiber. Opt Lett 22(13):961–963

Dudley JM, Coen S (2006) Supercontinuum generation in photonic crystal fiber. Rev Mod Phys 78(4):1135–1184

Südmeyer T, Brunner F, Innerhofer E, Paschotta R, Furusawa K, Baggett JC, Monro TM, Richardson DJ, Keller U (2003) Nonlinear femtosecond pulse compression at high average power levels by use of a large-mode-area holey fiber. Opt lett 28(20):1951–1953

Robinson J, Haworth C, Teng H, Smith R, Marangos J, Tisch J (2006) The generation of intense, transform-limited laser pulses with tunable duration from 6 to 30 fs in a differentially pumped hollow fibre. Appl Phys B 85(4):525–529

Nisoli M, Stagira S, De Silvestri S, Svelto O, Sartania S, Cheng Z, Lenzner M, Spielmann C, Krausz F (1997) A novel-high energy pulse compression system: generation of multigigawatt sub-5-fs pulses. Appl Phys B: Lasers Opt 65(2):189–196

Nisoli M, De Silvestri S, Svelto O (1996) Generation of high energy 10 fs pulses by a new pulse compression technique. Appl Phys Lett 68(20):2793–2795

Wei Z, Hao T, Chen-Xia Y, Xin Z, Xun H, Zhi-Yi W (2010) Generation of sub-2 cycle optical pulses with a differentially pumped hollow fiber. Chin. Phys Lett 27(5):0542111–0542114

Cavalieri AL, Goulielmakis E, Horvath B, Helml W, Schultze M, Fieß M, Pervak V, Veisz L, Yakovlev VS, Uiberacker M, Apolonski A, Krausz F, Kienberger R (2007) Intense 1.5-cycle near infrared laser waveforms and their use for the generation of ultra-broadband soft-x-ray harmonic continua. New J Phys 9(7):2421–2429

Minkovski N, Petrov GI, Saltiel SM, Albert O, Etchepare J (2004) Nonlinear polarization rotation and orthogonal polarization generation experienced in a single-beam configuration. J Opt Soc Am B 21(9):1659–1664

Jullien A, Albert O, Burgy F, Hamoniaux G, Rousseau J-P, Chambaret J-P, Augé-Rochereau F, Chériaux G, Etchepare J, Minkovski N, Saltiel SM (2005) 10-10 temporal contrast for femtosecond ultraintense lasers by cross-polarized wave generation. Optics Lett 30(8):920–922

Taha TR, Ablowitz MI (1984) Analytical and numerical aspects of certain nonlinear evolution equations. II. Numerical, nonlinear Schrödinger equation. J Comput Phys 55(2):203–230

Jullien A, Albert O, Chériaux G, Etchepare J, Kourtev S, Minkovski N, Saltiel SM (2006) A two crystal arrangement to fight efficiency saturation in cross-polarized wave generation. Opt Express 14(7):2760–2769

Lenzner M, Krüger J, Sartania S, Cheng Z, Spielmann Ch, Mourou G, Kautek W, Krausz F (1998) Femtosecond optical breakdown in dielectrics. Phys Rev Lett 80(18):4076–4079

Schwarz A (2014) Few-cycle phase-stable infrared OPCPA. PhD thesis

Bakker HJ, Cho GC, Kurz H, Wu Q, Zhang X-C (1998) Distortion of terahertz pulses in electro-optic sampling. J Opt Soc Am B 15(6), 1795–1801

Kato K (1986) Second-harmonic generation to 2048 A in beta-BaB_2O_4. IEEE J Quantum Electron 22(7):1013–1014

Brabec T, Krausz F (1997) Nonlinear optical pulse propagation in the single-cycle regime. Phys Rev Lett 78(17):3282–3285

Manzoni C, Cerullo G, De Silvestri S (2004) Ultrabroadband self-phase-stabilized pulses by difference-frequency generation. Opt Lett 29(22):2668–2670

Moses J, Suchowski H, Kärtner FX (2012) Fully efficient adiabatic frequency conversion of broadband Ti:sapphire oscillator pulses. Opt lett 37(9):1589–1591

Homann C, Bradler M, Förster M, Hommelhoff P, Riedle E (2012) Carrier-envelope phase stable sub-two-cycle pulses tunable around 1.8 μm at 100 kHz. Opt lett 37(10):1673–1675

Doumy G, Quéré F, Gobert O, Perdrix M, Martin P (2004) Complete characterization of a plasma mirror for the production of high-contrast ultraintense laser pulses. Phys Rev E 69(2):026402–026414

Renault A, Augé-Rochereau F, Planchon T, D'Oliveira P, Auguste T, Chériaux G, Chambaret J-P (2005) ASE contrast improvement with a non-linear filtering Sagnac interferometer. Opt Commun 248(4-6):535–541

Ricci A, Silva F, Jullien A, Cousin SL, Austin DR, Biegert J, Lopez-Martens R (2013) Generation of high-fidelity few-cycle pulses at 2.1 mm via cross-polarized wave generation. Opt Express 21(8):9711–9721

Liu J, Kobayashi T (2010) Generation and amplification of tunable multicolored femtosecond laser pulses by using cascaded four-wave mixing in transparent bulk media. Sensors 10(5):4296–4341 (Basel, Switzerland)

Cotel A, Jullien A, Forget N, Albert O, Chériaux G, Le Blanc C (2006) Nonlinear temporal pulse cleaning of a 1-μm optical parametric chirped-pulse amplification system. Appl Phys B 83(1):7–10

Fattahi H, Schwarz A, Keiber S, Karpowicz N (2013) Efficient, octave-spanning difference-frequency generation using few-cycle pulses in simple collinear geometry. Opt lett 38(20):4216–4219

Canova L, Kourtev S, Minkovski N, Jullien A, Lopez-Martens R, Albert O, Saltiel SM (2008) Efficient generation of cross-polarized femtosecond pulses in cubic crystals with holographic cut orientation. Appl Phys Lett 92(23):231102

Newport Corporation (2014)

Chapter 4
Regenerative Amplifiers

4.1 Yb:YAG Thin-Disk Regenerative Amplifier

Although[1] the construction of high energy laser systems is beyond the scope of this thesis, OPCPA requires a pump laser, whose characteristics form the ultimate limits on the possible specifications. In order to design and create an effective OPCPA, a thorough understanding of the underlying laser technology and its limits is required. The work in this thesis focuses on the use of Yb:YAG-based thin-disk regenerative amplifiers, e.g. (Metzger et al. 2009), as a useful source of high-energy, high power picosecond pulse trains to pump OPCPA systems, and the relevant considerations that must be made in order to fully exploit their capabilities.

In a typical laser gain medium, pump photons at the resonance frequency of the host material excite electrons to an upper atomic level, which subsequently decay to a lower excited state. Interaction of the seed photons with the excited electrons stimulates their decay to a lower energy level (in the ground state manifold) and consequently, photons with the same direction, phase, polarization and frequency of the seed photons are emitted.

In an ideal laser, where the medium consists of infinite numbers of electrons and in the absence of vacuum fluctuations, the amplified energy versus the input pump energy scales linearly. In the presence of vacuum fluctuations, amplification of the spontaneous emission depletes the medium and leaves no opportunity for the amplification of an injected pulse. In a finite medium, due to the restricted amount of atoms in the gain medium and therefore geometry-suppressed spontaneous emission, amplification can take place but still the depletion of the ground state can cause saturation and limit the amplification. Gain media with long upper state lifetimes have the potential for substantial population inversion under continuous pumping and are desirable for the construction of high-energy laser sources.

---

[1]This chapter is adopted from (Fattahi et al. 2014b).

H. Fattahi, *Third-Generation Femtosecond Technology*,
Springer Theses, DOI 10.1007/978-3-319-20025-5_4

Ytterbium-doped materials have a long upper state lifetime, broad emission, small density of quantum defects (Hönninger et al. 1999) and their broad absorption spectrum in the region accessible by efficient laser diodes makes them the most promising gain medium for amplifiers in disk (Metzger et al. 2009; Teisset et al. 2013), slab (Schulz et al. 2011) or fiber (Jauregui et al. 2013) geometries. However, in continuously pumped regenerative amplifiers, chaotic pulse train dynamics at repetition rates comparable to the inverse of the gain relaxation time restrict the operating parameters of the laser system (Dörring et al. 2004; Grishin et al. 2009). In this chapter, different approaches to overcome these instabilities in order to achieve reproducible, stable pulses with excellent temporal shape, are discussed.

Regenerative amplifiers operating at repetition rates close to the inverse of the inversion build-up time in the gain medium show special amplification dynamics. Here the amplification of a pulse in the cavity is strongly coupled to the previous pulse and can result in stable regimes of operation such as period doubling (bifurcation), tripling, or amplification of pulses with extremely random variations in pulse-to-pulse energy (chaotic regime). To visualize the process, let us consider a pulse entering the cavity and depleting the gain medium. If a second pulse enters the cavity when the population inversion of the gain medium is not at its maximum, then the amplification gain is less.

In the bifurcation regime, the third pulse enters the cavity when the population inversion has fully recovered and therefore carries away a large amount of energy and leaves the gain medium depleted. This cycle repeats and every second pulse gets amplified to the maximum energy. In the chaotic regime, the entrance of the third pulse and subsequent pulses into the cavity are not synchronized with the full recovery of the depleted gain medium and results in random pulse-to-pulse energy. Normal operation of the laser, where every pulse exiting the cavity has the same pulse energy, can be ensured by increasing the number of round trips (passes through the gain medium), such that at a given seed and pump power level, each pulse depletes the gain medium entirely. Increasing the number of round trips increases the number of passes through the Pockels cell (PC) and air. This increases the accumulated nonlinear phase on the amplified pulses and initiates nonlinear effects like SPM which consequently influence the temporal and spectral shape of the pulse.

The induced nonlinear phase is described as:

$$\frac{\mathrm{d}\Phi_{nl}(z,t)}{\mathrm{d}z} = -\frac{\omega}{c} n_2 I(t,z). \tag{4.1}$$

This instantaneous phase adds a nonlinear phase to the pulse. After propagating distance L inside the PC and air, the accumulated nonlinear phase in terms of the B-integral is:

$$B(t,r) = \frac{2\pi}{\lambda_{vac}} \int_0^L n_2 I(t,r,z) dz, \tag{4.2}$$

where r is the transverse coordinate of the laser beam. The spatial variation of the B-integral implies that perfect compensation of the induced nonlinear phase is not possible. The radial variation also forms a nonlinear lens inside the cavity which can cause coupling to higher order modes. These modes are filtered out by the cavity but cause reduction in the efficiency as they increase cavity losses.

B-integral values of ~ 2 could double the pulse duration (Chuang et al. 1993). Even for the modest value of $B \leq 1$ significant energy can be removed form the central peak of the pulse and appear as a pedestal, decreasing the contrast and lowering the peak power (Perry et al. 1994). In the SPM process the interference between the newly generated frequencies and the spectrum of the pulse causes a change in the spectral bandwidth of the pulse. In Fig. 4.1 the spectral and temporal effects for different chirp values applied to the pulse are shown. For negatively chirped pulses with a GDD of -0.27×10^7 fs^2, destructive interference between the newly generated

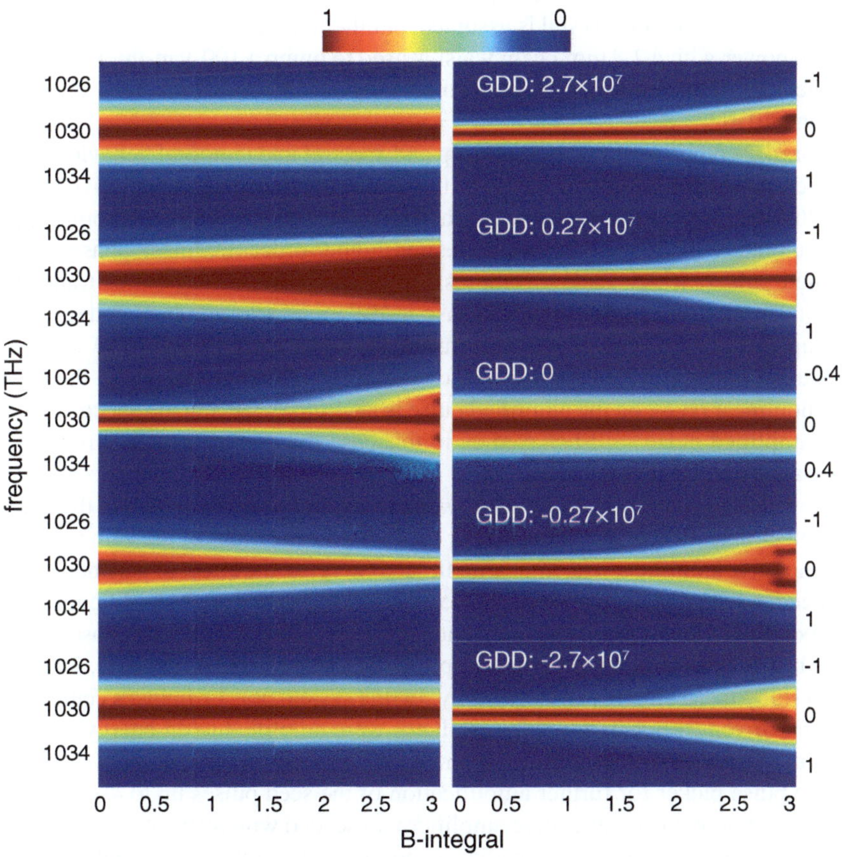

Fig. 4.1 Spectral (*right panels*) and temporal (*left panels*) evolution of a pulse, stretched with different amounts of dispersion versus B-integral

photons in SPM and the spectrum of the pulse causes spectral narrowing. For heavily chirped pulses, however, SPM primarily influences the spectral phase of the pulse, leaving the shape and the bandwidth of the spectrum unchanged.

In regenerative amplifiers with low input seed energy, a large number of round trips are required to reach energy saturation. To study the effect of B-integral in low-energy-seeded regenerative amplifiers, a simple 1-D model (Appendix B) based on the rate equations for continuously-pumped Yb:YAG (Antognini et al. 2009) is used. The model includes the wavelength-dependent pulse energy buildup for the full pulse train and the nonlinearity acting on the amplified pulses caused by the PC and air is calculated in each round trip. The gain in each pass is obtained via the temperature-dependent emission cross-section according to (Chen et al. 2003).

In the simulation a seed pulse with 10 nm spectral bandwidth is stretched using GDD of $1.19 \times 10^8\,\mathrm{fs}^2$, to approximately 2 ns. The pulse travels for 40 ns in the cavity in each round trip. A 20-mm-long PC (traversed twice per round trip) with a nonlinearity of $2.99 \times 10^{-20}\,\mathrm{cm}^2/\mathrm{W}$ (Desalvo et al. 1996) and a beam waist of 3.5 mm ($1/e^2$) on the PC crystal is assumed, similar to (Metzger et al. 2009). 225 W of pump power with a 2.8 mm beam waist is used to pump a 100-μm-thick Yb:YAG gain medium. The simulation is performed for two different input seed energies, 80 pJ and 40 nJ.

Figure 4.2 compares the pulse build-up and spectral and temporal behavior of the amplified pulses for these two cases. In the low-energy-seeded regenerative amplifier, 126 round trips are required to reach saturation. At this number of round trips, the accumulated B-integral causes undesired higher-order nonlinear phase on the amplified pulses, which makes compression of the pulse to its Fourier transform limit challenging. Any additional increase in the round trip numbers causes pulse break-up in the time domain. By increasing the input seed energy level to 40 nJ, the required number of round trips to reach saturation reduces to 90. Here, the introduced nonlinear phase on the amplified pulse is negligible, leaving the spectral phase of the amplified pulse well-behaved as in this case the nonlinear path through the regenerative amplifier is reduced.

In Fig. 4.2, for the case of low-energy-seed, spectral narrowing versus the round trip number is visible, which is caused by the limited bandwidth of the gain medium. In regenerative amplifiers with higher input seed energies, the effect of gain narrowing is reduced, which results in a longer pulse inside of the cavity after amplification and correspondingly lower pulse intensity and less accumulated nonlinear phase.

A Yb:YAG thin-disk regenerative amplifier described in (Metzger et al. 2009), is used to study these effects experimentally. In this system, the 1 pJ pulses centered at 1030 nm, driven from a Ti:sapphire oscillator were amplified to an energy of 80 pJ with a diode-pumped fiber pre-amplifier. A second fiber pre-amplifier was placed after the pulse picker for further amplification of the seed pulses up to 40 nJ. In the first measurement, the regenerative amplifier was seeded with 80 pJ of energy. Using 107 cavity round trips, after intermediate chaos, period doubling and a stable regime of operation was achieved. The amplified pulses were passed through a grating pair for temporal compression and characterized by SH-FROG exploiting a 10 μm BBO crystal as the nonlinear medium.

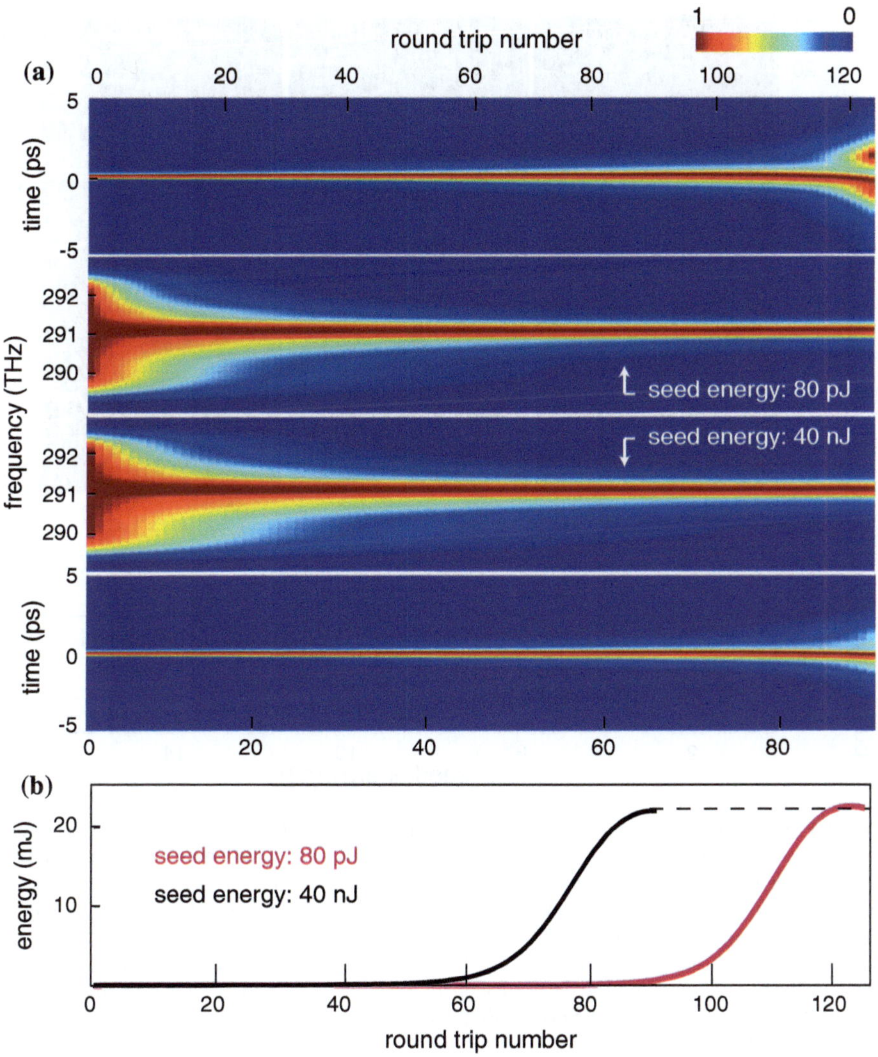

Fig. 4.2 **a** Simulated temporal and spectral profile of the amplified pulses versus round trip number in a regenerative amplifier for two different input seed energies: 80 pJ and 40 nJ. For the case of low-energy-seed, the accumulated nonlinear phase, causes undesired temporal features on the amplified pulses (The temporal and spectral amplitudes in all cases are normalized to unity). **b** Pulse build-up in the regenerative amplifier for the two different input seed energies. For the case of the higher seed energy, saturation happens after a lower number of round trips

A stable regime of operation with no chaotic behavior or bifurcation was observed after increasing the input seed energy to the cavity (Grishin et al. 2007). Figure 4.3 shows the energy distribution in a bifurcation diagram acquired by incrementing the pump power over the range of 50–250 W with the number of cavity round trips kept

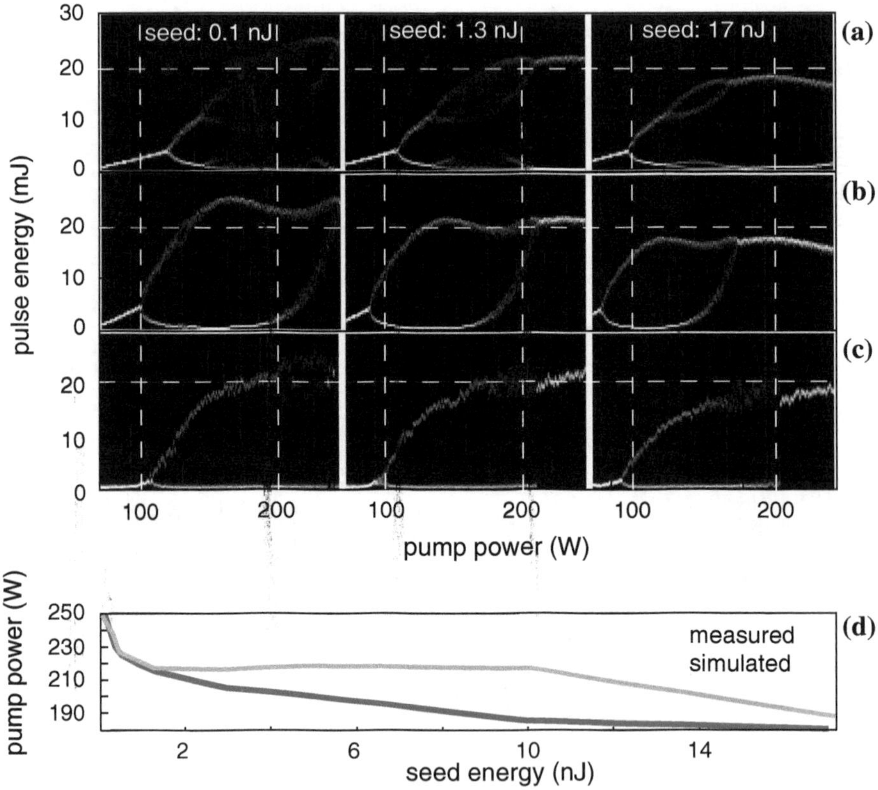

Fig. 4.3 Simulated and measured amplification dynamics for CW-pumped Yb:YAG thin-disk regenerative amplifier operating at different repetition rates and seed energies. **a** The amplifier with a PC driven at 6 kHz repetition rate enters a chaotic regime for low seed energy and reaches stable bifurcation at higher pump power. Increasing the input seed energy reduces the chaos in the cavity. **b** The same system with a PC driven at 3 kHz repetition rate. Here no chaotic regime is observed and the bifurcation is removed at a higher pump power. The pump power difference for reaching the bifurcation point and the normal operation point is reduced by increasing the input seed energy. **c** Measured dynamics at 3 kHz. **d** Required pump power for stable operation as a function of pulse energy at 3 kHz

constant at 70. Here the PC was driven at 3 kHz repetition rate. For the input seed energy of 0.1 nJ, the bifurcation happened at 100 W of pump power. By increasing the pump power to 250 W, the amplifier went out of the doubling regime and the amplification happened at 3 kHz repetition rate. By increasing the input seed energy the bifurcation and normal amplification points were shifted to lower pump powers.

The simulated and measured data are in good agreement. In both cases, there is no sign of chaotic behavior for high energy seeding. Figure 4.3a shows the simulated data in the case of 6 kHz operation similar to (Metzger et al. 2009). Here, the bifurcation point is shifted to a higher pump power compared to 3 kHz operation. After period doubling, the amplifier goes to the chaotic regime and at high pump power reaches a stable doubling regime. The separation of this point and the chaotic zone is determined by the input seed energy level. Running the amplifier with a PC driven at 3 kHz is advantageous as it simplifies synchronization with OPCPA seed sources and reduces the thermal load on the PC crystal. Due to hysteresis, the amplifier has the tendency to stay in the normal operation regime. After emerging from bifurcation, even for slightly reduced pump power, the laser stays in the normal operation mode as the cavity design is optimized for high-power operation. Therefore the cavity efficiency increases slightly as the system enters the normal amplification mode and the output power doubles. The hysteresis and thermal effects which are not included in the simulation are the primary source of discrepancy between the measured and simulated data.

Figure 4.4b shows the effect of the cavity dynamics on the temporal profile of the compressed pulses. The characterization of the unsaturated amplified pulses at 14 W seeded with 80 pJ does not show any phase distortion and the pulses are nearly compressed to their Fourier-transform limit (Fig. 4.4a). However, after running the laser in the saturated regime with an output power of 58 W, SPM begins to distort the temporal phase of the amplified pulses, although the shape of the spectrum is not affected. By increasing the input seed energy to 300 pJ, increasing the pump power and reducing the cavity round trip number by 10, the characterized 61 W, compressed pulses show a good temporal profile indicating that the effect of SPM in this case in negligible (Fig. 4.4c). Figure 4.4d shows calculated nonlinear group delay distortion of the amplified pulses for varying input seed energies to the amplifier, explaining the temporal behavior of the measured pulse in Fig. 4.4b. The FROG G-error of the reconstructed trace in Fig. 4.4b is 0.022 which is more than the the G-error of Fig. 4.4c, 0.0087. This could be an indication of pulse-to-pulse variations in temporal profile (Ratner et al. 2012).

Different approaches can be used to eliminate the undesired SPM effects in amplifiers while maintaining stable operation. Table 4.1 summarises different approaches with their pros and cons.

Effect of Suboptimal Amplifier Design in Subsequent Nonlinear processes In the presence of significant SPM in the system, small changes in the seed energy level and cavity alignment result in large variations of the amplified output pulses. For an OPCPA pumped by such pulses these variations affect the efficiency of the amplified pulses in the subsequent broadband OPCPA stages. For a regenerative amplifier operating at the border of nonlinearity, variations of the spectral phase to the evolution of the pulse inside of the cavity can result in large fluctuations in the OPCPA output, although the pulse-to-pulse energy fluctuations of the pulses of the saturated amplifier are at their minimum. To study these effects, the spectrum and gain of an OPCPA system are simulated.

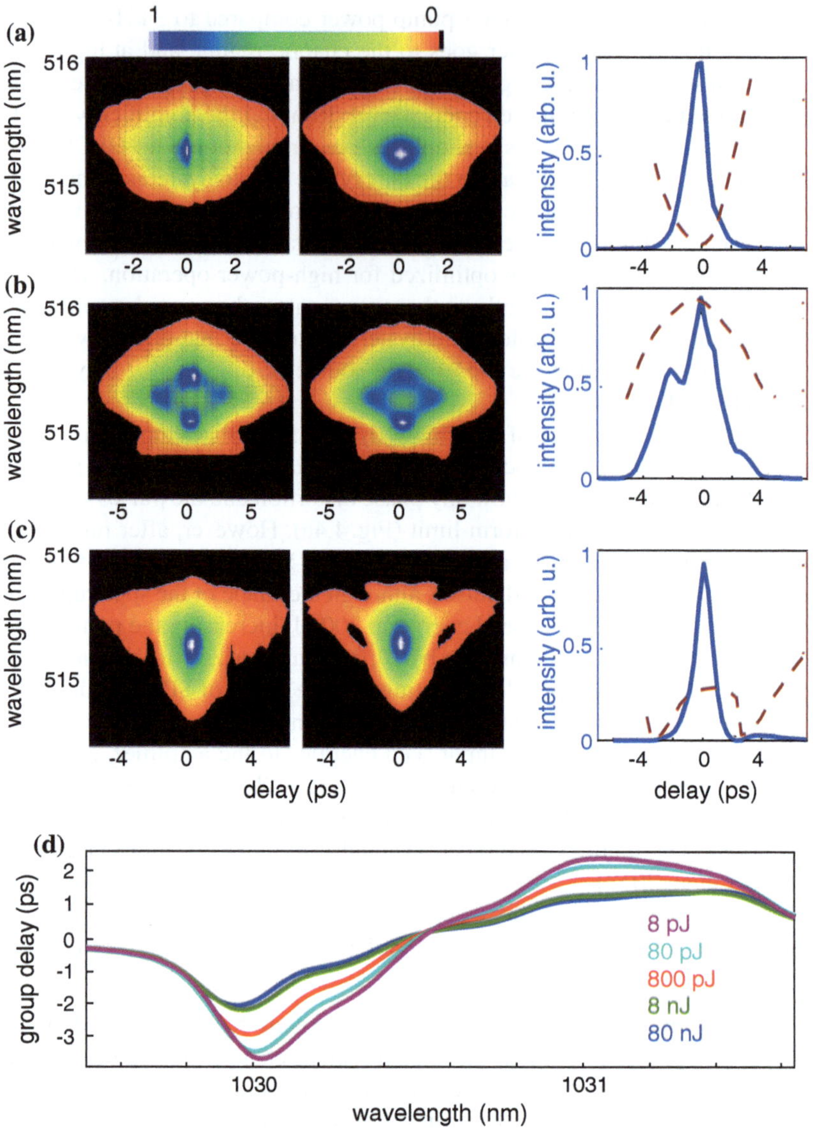

Fig. 4.4 **a** Measured SH-FROG traces of the 14 W amplified pulses in a regenerative amplifier seeded with 80 pJ of energy and making 107 cavity round trips (the laser operated in the unsaturated regime with unstable laser output). **b** Measured SH-FROG trace for the same pulses amplified to 58 W, showing strong SPM in the time domain. **c** Temporal profile of 300 pJ amplified pulses to 61 W with increased pump power and reduced round trip number. The effect of SPM in this case is negligible. **d** Calculated group delay resulting from the induced nonlinear phase on the amplified pulses with different input seed energies. The pump power for all cases was kept constant and saturation was ensured by changing the number of cavity round trips

Table 4.1 Possible approaches for the reduction of SPM in a thin-disk regenerative amplifier

Approach	Mechanisms	Negative consequences
Reduced round trips with higher pump power	• Reduced nonlinear path length • Reduced cavity losses leading to less gain narrowing at same energy	• Increased thermal load • Smaller range of stable operating power
Increase seed level viafurther broadband pre-amplification	• Reduced gain narrowing • Reduced nonlinear propagation distance via reduced round trip number	• Increased system complexity/cost • Longer seed beam path
Decrease pump spot size	• Faster inversion buildup for given pump power, allowing stable operation at lower power	• Increased thermal gradient in gain medium • Reduced saturation energy
Increase stretcher and compressor dispersion	• Reduced intensity at given pulse energy	• Increased cost and alignment sensitivity of compressor and stretcher • Additional timing jitter when synchronized with external seed
Increased beamsize in nonlinear cavity elements via reflective telescope	• If a single element (i.e. Pockels cell) is responsible, reduced intensity	• Increased complexity of the laser cavity • Requires large cavity optics
Operation in vacuum	• Reduced nonlinearity of free beam path	• Cost and complexity • More difficult thermal management

The pulses from Fig. 4.4b, c with an identical pulse energy of 3 mJ and a beam diameter of 1.2 mm at full width half maximum are used to amplify a broad spectral range in a single OPCPA stage consisting of 1 mm BBO crystal with a d_{eff} value of 2.3 pm/V, a phase-matching angle of 24.5°, and noncollinear angle of 2.7°, seeded with 1 μJ of energy. As shown in Fig. 4.5 this variation in pump pulses results in a factor of 3.5 difference in pump-signal efficiency and modifications in the amplified spectrum. Suboptimal pump pulses also affect the spectrum and efficiency of the generated second harmonic (Fig. 4.5c).

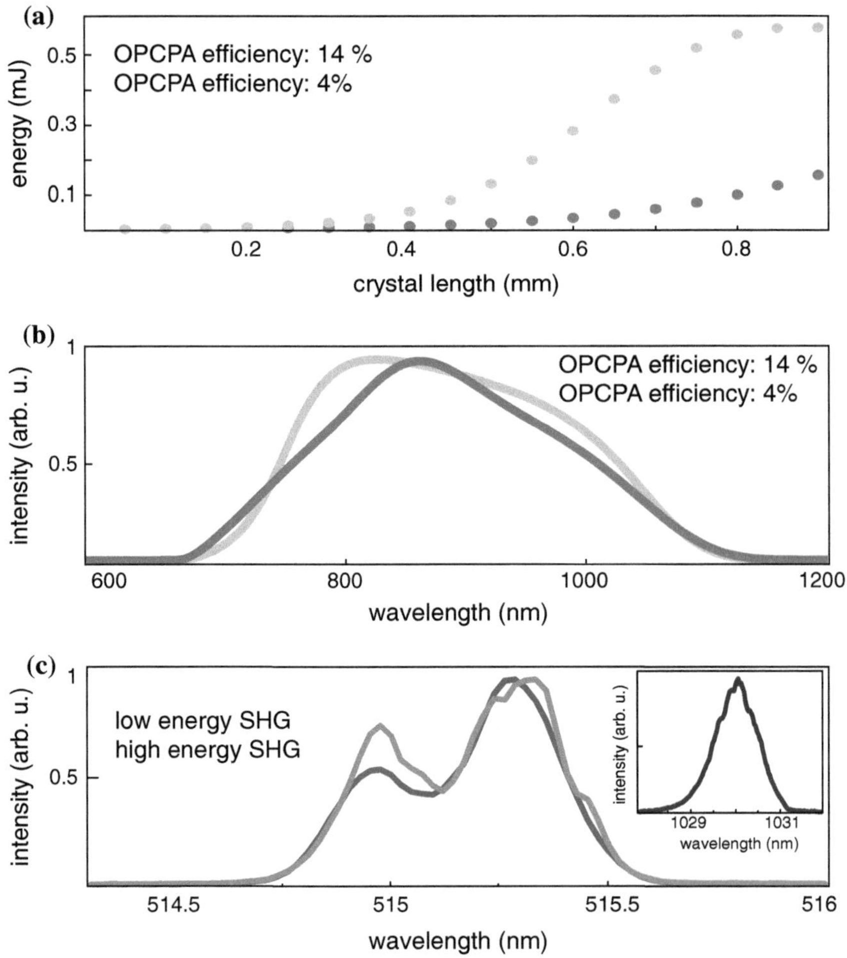

Fig. 4.5 Simulation of the effect of amplified SPM on the output of a broadband OPCPA. The orange curves are simulated with the pulse measured in Fig. 4.4c and the green curves correspond to the pulse in Fig. 4.4b. **a** Gain versus crystal length in the OPCPA, showing saturation with the SPM-free pulse, and a factor of 3.5 reduction in pulse energy with the distorted pump pulse. **b** Modification of the spectrum resulting from the pump temporal profile (normalized). **c** The measured second-harmonic spectra of the suboptimal pump pulses (Fig. 4.4b) in the saturated (high energy) and unsaturated (low energy) regime. The hole in the second harmonic spectra is caused by the distorted spectral phase of the input pump pulses. The pump spectrum is shown in the *inset*

Conclusion Stable OPCPA demands not only excellent pump energy stability, but also a high degree of reproducibility of the temporal shape of the pump pulses, which can only be ensured by careful matching of the amplifier design to the seed specifications. The presented simple one-dimensional simulation and the measurements show the chaotic regime in the ytterbium-doped regenerative amplifiers is avoidable by increasing the input seed energy. Here the accumulated nonlinear phase in the cavity is minimized and a reproducible spectral phase is achievable. The effect of a suboptimial amplifier on a subsequent nonlinear stage (here OPCPA) is numerically studied, which emphases the high demands on well-behaved phase of pump pulses in an OPCPA system. By ensuring that these effects are well controlled, thin-disk regenerative amplifiers are an ideal route to stable and efficient pumping of high-power, high-energy, broadband OPCPAs.

4.2 Second Harmonic Generation

Pumping OPCPAs in the near infrared and visible requires the generation of low-order harmonics of the near-infrared emission of the Yb:YAG medium (Sect. 2.2). Although there are numerous options for second harmonic generation, critically phase-matched LBO and BBO are among the best candidates for low-order harmonic generation of near-1-ps pulses at 1030 nm. They have fairly high nonlinear coefficients and damage thresholds. BBO has a higher nonlinearity than LBO but a larger spatial walk-off and is limited in the available aperture, as summarized in Fig. 4.6a.

Figure 4.6b shows the simulated SHG gain versus crystal length for 1.6 ps pulses centered at 1030 nm with a pulse energy of 200 mJ and a beam diameter of 10 mm in type I BBO and LBO crystals. In the LBO crystal with a crystal thickness twice

(a)

	n_2 (cm²/W)	walk-off	available size	α (cm⁻¹)	Type I	Type II
BBO	7.4×10^{-20}	55.3 mrad (I) 68.47 mrad (II)	25×25 mm²	0.01 @ 532	Θ= 23.4	Θ= 33.7
LBO	1.9×10^{-20}	8.16 mrad	70×70 mm²	0.00035 @1064 nm	Φ= 13.77°	Θ=25.4°

(b)

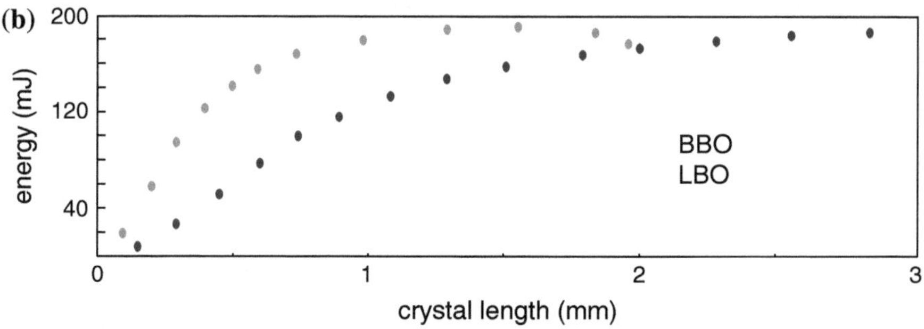

Fig. 4.6 **a** Parameters of BBO and LBO crystal for SHG. **b** SHG gain versus crystal length of a 200 mJ, 1.6 ps pulses in a type I, BBO and LBO crystal from simulation. α values in the table indicate the absorption coefficient of each crystal

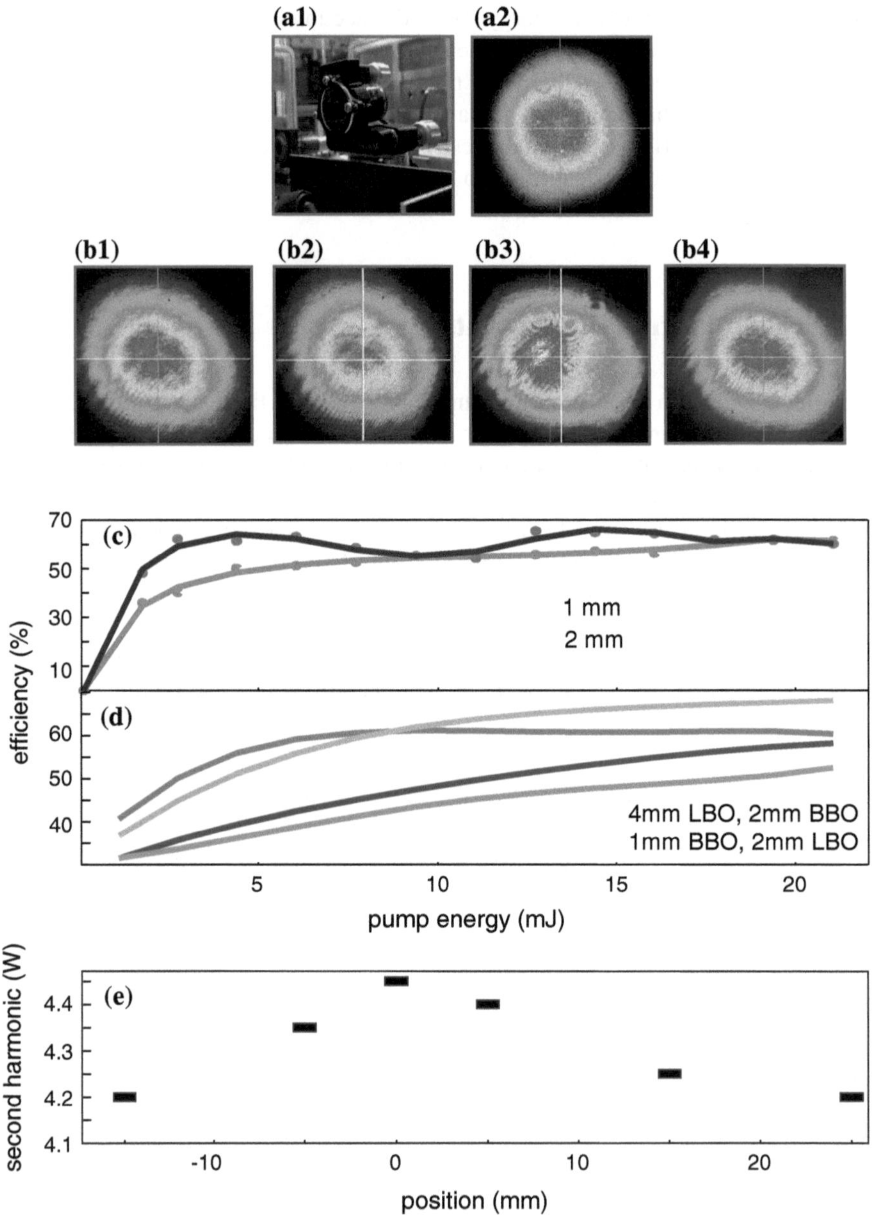

Fig. 4.7 **a, b** Home-made passive adaptive mirror used before the grating compressor of the disk laser (**a1**). The beam profile of the amplifier after the adaptive mirror and the cylindrical telescope (**a2**). The beam profile of the generated second harmonic in a 1 mm BBO (**b1**), 2 mm LBO (**b2**), 2 mm BBO (**b3**) and 4 mm LBO (**b4**) crystals. **c** SHG in a 1 mm and 2 mm BBO crystal with the intensity of 120 GW/cm^2. **d** SHG in a 1 mm and 2 mm BBO and 2 mm and 4 mm LBO crystals for a pump intensity of 70 GW/cm^2. **e** Second harmonic signal versus grating separation in a 1 mm BBO crystal

that of the BBO crystal, the generated second harmonic reaches saturation. Due to reduced spatial walk-off between the second harmonic beam and the fundamental beam in LBO, a conversion efficiency similar to BBO can be achieved, despite having a smaller d_{eff} value than BBO. The calculated conversion efficiency does not reach 100 % becuase of phase-mismatch between the second harmonic and the pump beam, and the temporal and spatial walk-off in the nonlinear crystals.

Experiment The discussed Yb:YAG thin-disk regenerative amplifier was used to study SHG in LBO and BBO crystals. The output beam profile of the laser after compression became elliptical as the grating pair used for the pulse compression was not entirely flat. The ellipticity was compensated by implementing a home-made adaptive mirror before the grating compressor. By tuning the curvature of the thin, flat mirror the beam ellipticity was improved to 0.78 and the residual ellipticity was further improved to 0.83 by using a set of cylindrical mirrors with radii of curvature of −1574 and 500 mm (Fig. 4.7).

A refelective telescope consisting of mirrors with radii of curvature of −2000 and 800 mm was used to adjust the beam size on the crystal in order to reach 120 GW/cm^2 intensity, assuming a Gaussian beam in time and space. The SHG in a 2 mm BBO crystal reached 70 % conversion efficiency and saturated at 5 mJ of input pump energy. By increasing the input pump energy, the energy backconversion from the SHG signal to the pump resulted in oscillatory behavior in the efficiency curve and degradation of the spatial profile of the generated second harmonic (Fig. 4.7b.3). The generated second harmonic in a 1 mm BBO crystal with the same pump parameters reached saturation while the spatial profile of the pulses preserved its quality. A pump intensity of 70 GW/cm^2 on a 4 mm LBO crystal resulted in an efficiency of 70 % with an excellent beam quality. SHG could also be used as a measure for fine tuning the grating pair compressor. Figure 4.7e) shows the sensitivity of the SHG conversion efficiency to the temporal profile of the pump pulses.

Conclusion As it is shown in this section, employing BBO or LBO crystals for frequency doubling of a 20 mJ, 1.6 ps pulses of an Yb:YAG amplifier, leads to pump-to-second harmonic conversion efficiency above 65 %. This high efficiency indicates the optimum temporal and spatial properties of the amplifier pulses. High conversion efficiencies in both crystals is achieved, even though the nonlinearity of a LBO crystal is about half of the BBO crystal, due to the lower temporal walk-off in a LBO.

References

Antognini A, Schuhmann K, Amaro FD, Biraben F, Dax A, Giesen A, Graf T, Hänsch TW, Indelicato P, Julien L, Kao C-Y, Knowles P-E, Kottmann F, Le Bigot E, Liu Y-W, Ludhova L, Moschüring N, Mulhauser F, Nebel T, Nez F, Rabinowitz P, Schwob C, Taqqu D, Pohl R (2009) Thin-disk Yb:YAG oscillator-amplifier laser, ASE, and effective Yb:YAG lifetime. IEEE J Quantum Electron 45(8):993–1005

Chen B, Dong J, Patel M, Chen Y, Kar A, Bass MA (2003) Modeling of high-power solid-state slab lasers. In: Scheps R (ed) SPIE proceeding, pp 1–10

Chuang Y-H, Zheng L, Meyerhofer D (1993) Propagation of light pulses in a chirped-pulse-amplification laser. IEEE J Quantum Electron 29(1):270–280

Desalvo R, Said AA, Hagan DJ, Stryland EWV, Member S, Sheik-bahae M (1996) Infrared to ultra-violet measurements of two-photon absorption and n2 in wide bandgap solids. IEEE J Quantum Electron 32(8):1324–1333

Dörring J, Killi A, Morgner U, Lang A, Lederer M, Kopf D (2004) Period doubling and deterministic chaos in continuously pumped regenerative amplifiers. Opt Express 12(8):1759–1768

Fattahi H, Schwarz A, Geng XT, Keiber S, Kim DE, Krausz F, Karpowicz N (2014b) Decoupling chaotic amplification and nonlinear phase in high-energy thin-disk amplifiers for stable OPCPA pumping. Opt Express 22(25):31440–31447

Grishin M, Gulbinas V, Michailovas A (2007) Dynamics of high repetition rate regenerative. Opt Express 15(15):9434–9443

Grishin M, Gulbinas V, Michailovas A (2009) Bifurcation suppression for stability improvement in Nd:YVO4 regenerative amplifier. Opt Express 17(18):15700–15708

Hönninger C, Paschotta R, Graf M, Morier-Genoud F, Zhang G, Moser M, Biswal S, Nees J, Braun A, Mourou GA, Johannsen I, Giesen A, Seeber W, Keller U (1999) Ultrafast ytterbium-doped bulk lasers and laser amplifiers. Appl Phys B: Lasers Opt 69(1):3–17

Jauregui C, Limpert J, Tünnermann A (2013) High-power fibre lasers. Nat Photonics 7(11):861–867

Metzger T, Schwarz A, Teisset CY, Sutter D, Killi A, Kienberger R, Krausz F (2009) High-repetition-rate picosecond pump laser based on a Yb:YAG disk amplifier for optical parametric amplification. Opt Lett 34(14):2123–2125

Perry MD, Ditmire T, Stuart BC (1994) Self-phase modulation in chirped-pulse amplification. Opt Lett 19(24):2149–2151

Ratner J, Steinmeyer G, Wong TC, Bartels R, Trebino R (2012) Coherent artifact in modern pulse measurements. Opt Lett 37(14):2874–2876

Schulz M, Riedel R, Willner A, Mans T, Schnitzler C, Russbueldt P, Dolkemeyer J, Seise E, Gottschall T, Hädrich S, Duesterer S, Schlarb H, Feldhaus J, Limpert J, Faatz B, Tünnermann A, Rossbach J, Drescher M, Tavella F (2011) Yb:YAG Innoslab amplifier: efficient high repetition rate subpicosecond pumping system for optical parametric chirped pulse amplification. Opt lett 36(13):2456–2458

Teisset C, Schultze M, Bessing R, Haefner M, Prinz S, Sutter D, Metzger T (2013) 300 W picosecond thin-disk regenerative amplifier at 10 kHz repetition rate. In: Advanced solid-state lasers congress postdeadline, p JTh5A.1

Chapter 5
Experimental OPCPA

In this chapter, broadband seed pulses demonstrated in Chap. 3 and the turn key Yb:YAG pump laser discussed in Chap. 4 are employed to develop OPCPA systems with particularly high conversion efficiency.

Conversion efficiency in an OPCPA is limited by the group velocity mismatch between pump, signal and idler during the amplification process, deficient spatial overlap between the interacting beams caused by spatial walk-off in the nonlinear crystal, the temporal and spatial profiles of the seed and pump beams, and finally, the amplification of superfluorescence.

Overcoming these limitations to increase the efficiency has been the subject of several previous studies. By using a spatiotemporally shaped pump pulse, a pump-to-signal conversion efficiency as high as 40 % (theoretically) (Guardalben et al. 2003) and 25 % (Chekhlov et al. 2006), 29 % (Waxer et al. 2003) and 34 % (Bagnoud et al. 2005) (experimentally) have been reported. However, there is no straightforward possibility for spatiotemporal shape of the ps-pump laser pulses. Also, the amplified spectrum of the mentioned OPCPAs is narrow compared to the bandwidth of few-cycle pulses. For 100-ps-pumped few-cycle OPCPAs, a 27 % conversion efficiency was demonstrated by applying a complex, double-pass OPCPA configuration (Ishii et al. 2005) and in a near-1 ps-pumped OPCPA system, few-cycle pulses with 25 % pump-to-signal efficiency were shown (Fattahi et al. 2012b).

In this chapter, the theoretical analysis and experimental implementation of three different techniques for the simultaneous optimization of conversion efficiency and amplified spectral bandwidth in near-1 ps-pumped OPCPA systems are explored.

H. Fattahi, *Third-Generation Femtosecond Technology*,
Springer Theses, DOI 10.1007/978-3-319-20025-5_5

(a)

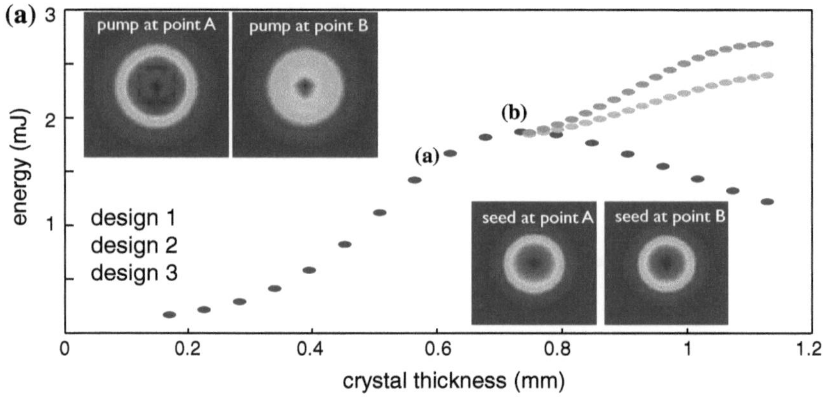

(b)

	Design 1	Design 2		Design 3	
	1st stage	1st stage	2nd stage	1st stage	2nd stage
L_c (mm)	1.13	0.68	0.45	0.68	0.45
$E_{s\text{-amp}}$ (mJ)	1.2	1.8	2.4	1.8	2.7
Φ_p (mm)	0.95	0.95	–	0.95	0.75
Φ_s (mm)	0.95	0.95	–	0.95	D^*_{amp}
efficiency (%)	15.75	24.3	13.7	24.3	21.3
total efficiency (%)	15.75	34.3		38.6	

D^*_{amp}:The diameter of the amplified beam

Fig. 5.1 **a** Simulated amplified signal energy over the crystal length for three different designs. Design 1 consists of one amplification stage. Designs 2 and 3 consist of two amplification stages, where the residual pulse energy after the first amplification stage is used to pump the second stage. In Design 3, the pump after the first amplification stage is resized to keep the pump intensity on the crystal similar to the first stage. The beam profiles of the pump after the first OPCPA stage at points (*A*) and (*B*) and their corresponding amplified-signal-beam profiles after the second OPCPA stage are shown in the *inset*. **b** The simulated parameters of the three different designs

5.1 Reusing the Pump

In order to increase the conversion efficiency of the near-1-ps-pumped OPCPAs, the residual of the pump energy from one stage can be reused in a subsequent OPCPA stage (Schultze et al. 2010).[1] Figure 5.1 compares three different simulated OPCPA designs.

In this calculation, a Gaussian structure for the pump in both space and time and a Gaussian structure for the seed in space is assumed. The fourth-order super-Gaussian seed spectrum is linearly chirped to a 1.6 ps pulse duration with the spectrum

[1] This section is adapted from (Fattahi et al. 2012a).

spanning from 600–1100 nm. The noncollinear type I geometry in a BBO crystal with an internal pump-signal angle of 2.7° and a phase-matching angle of 24.5° is used. The first design consists of a single OPCPA stage using a BBO crystal, pumped with an intensity of approximately 100 GW/cm^2. In the second design, firstly, a 0.68 mm-thick BBO crystal is pumped with 7 mJ and the residual pump energy is reused in a subsequent stage consisting of another 0.45 mm-thick BBO crystal. In this design a higher pump-to-signal conversion efficiency is expected due to the possibility that the noncollinear pump-signal angle and the phase-matching angle can be readjusted and optimized between the two stages.

The third configuration is a two-stage OPCPA system similar to the second design except that the pump-beam size between the two stages is readjusted to provide about 100 GW/cm^2 pump intensity on the second crystal in order to compensate for the reduced pump intensity after its depletion in the first stage. The amplified signal energy as a function of total crystal length for three different designs is shown in Fig. 5.1, where the third design results in a factor of 2.5 increase in the conversion efficiency of the OPCPA system.

The observed increase in efficiency for this scheme arises from the freedom to readjust and optimize the phase-matching angle, noncollinear pump-signal angle, and spatial and temporal overlap for the second crystal. Importantly, the idler in the second amplification stage is built up from zero energy, and therefore the amplification saturation happens in a longer total crystal length. Additionally, in the third design, the higher pump intensity on the last crystal results in a more efficient energy extraction compared to the second design. In Fig. 5.1a, the corresponding pump-beam profiles at two different points A and B on the amplification gain curve are shown. Although the pump-beam profile changes dramatically from point A to B, the amplified signal beams preserve their quality. Therefore, the third design was chosen for experimental realization.

Experiment The experimental OPCPA setup consisted of five main ingredients: the pump laser (see: Chap. 4), the Ti:sapphire oscillator and amplifier, followed by the broadband nonlinear seed generation (see: Sect. 3.2), the timing jitter compensation system (Schwarz et al. 2012), and three OPCPA stages followed by a chirped-mirror compressor.

52 W of the Yb:YAG regenerative amplifier output was frequency doubled in a 1 mm type I, BBO crystal with a conversion efficiency as high as 57 % (Fig. 5.2). The choice of the material and the length of the SHG crystal was based on minimizing the total B-integral in the whole OPCPA chain while maintaining a good pump-to-second-harmonic conversion efficiency.

A 4 mm-thick SF_{57} plate was used to temporally stretch the 10 μJ broadband OPCPA seed in order to provide good temporal overlap between the pump and signal pulses. The broadband seed was sent through the entire beam path in the absence of amplification consisting in total of 5.5 mm BBO from the OPCPA stages, 4 mm of SF_{57} glass and air so that the phase could be characterized by a multi-shot, cross-correlation frequency-resolved optical gating (XFROG) device incorporating a 20 μm BBO crystal cut at 29° as the nonlinear medium. A fraction of the Ti:sapphire

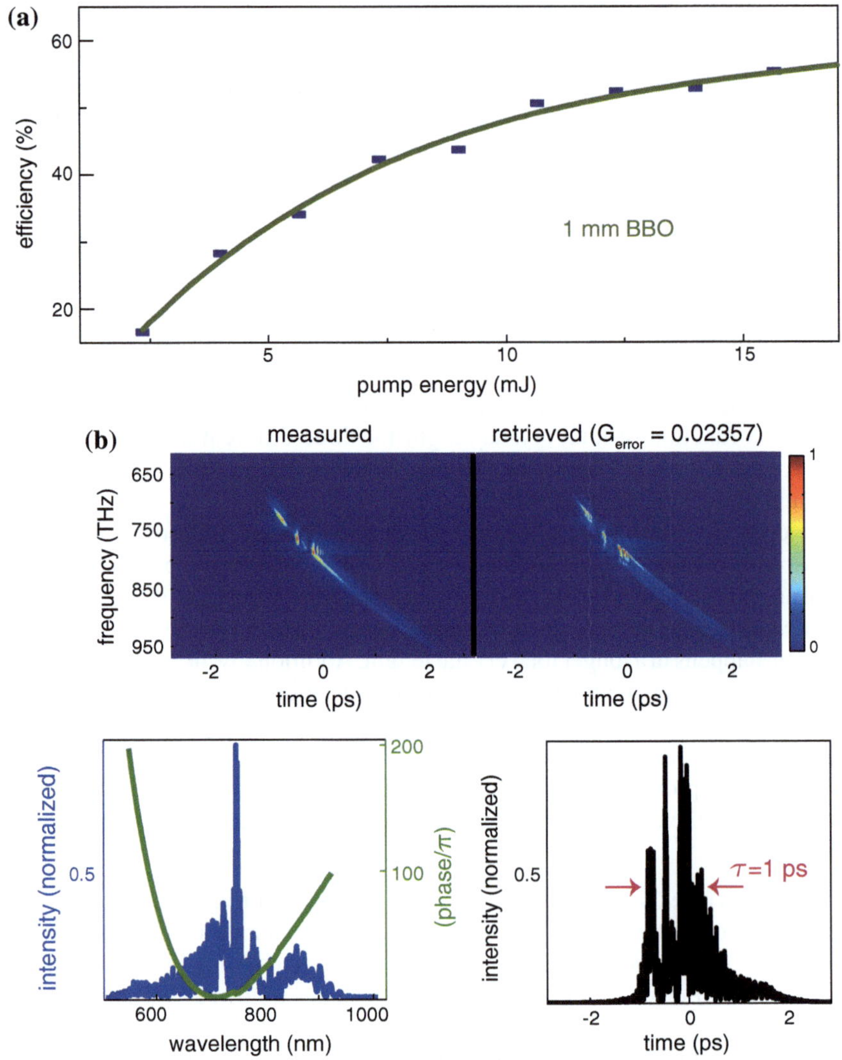

Fig. 5.2 **a** The efficiency curve of SHG in a 1 mm BBO crystal. **b** The XFROG measurement of the OPCPA signal passing through the entire beam path, consisting altogether 5.5 mm BBO and a 4 mm SF57. SF57 is used to temporally stretch the signal to ensure a proper temporal overlap between signal and pump pulses

amplifier output provided the reference pulse (see: Sect. 3.2). The pulse duration of the seed (assuming a Gaussian fit for the retrieved time structure) is 1.1 ps which ensures sufficiently good temporal matching between the seed and pump pulses in the OPCPA stages (Fig. 5.2). The XFROG measurement indicated that the spectral phase of the OPCPA seed was well-behaved and compressible.

The first OPCPA stage serves as a preamplifier for the low-energy broadband seed pulses. 1 mJ of the frequency doubled output of the thin-disk amplifier was used for pumping a 1.5 mm type I, BBO crystal and an output pulse energy of 120 μJ was achieved. Using a 3 mm BBO crystal in the first stage resulted in 800 mW of signal output, but this configuration resulted in a total of 400 mW of superfluorescence, measured after the third stage when the seed was blocked before the first amplification stage, while no further increase in the efficiency of the system was observed. Therefore the 1.5 mm BBO crystal was chosen for the first stage to minimize the superfluorescence after the third stage.

Up to 7.3 mJ of energy at 515 nm was used to pump the 2 mm BBO crystal at the second OPCPA stage. The remaining 5.6 mJ pump energy after the second stage preserved its good beam quality (Fig. 5.3), (except for the large ellipticity caused by the grating pair compressor of the amplifier, see Chap. 4). Subsequently, the size of the remaining pump beam was reduced to provide 80 GW/cm^2 pump intensity on a 2 mm BBO crystal in the third amplification stage. In the last two OPCPA stages an optical-to-optical conversion efficiency of 32 % was achieved (Fig. 5.4). No measurable superfluorescence background was observed when blocking the signal beam in front of the first stage.

The 350 nm-broad-amplified signal supports a Fourier transform limit pulse duration of 5.7 fs as shown in Fig. 5.5. Preliminary compression using a double-angle chirped-mirror compressor similar to (Fattahi et al. 2012b) resulted in a pulse

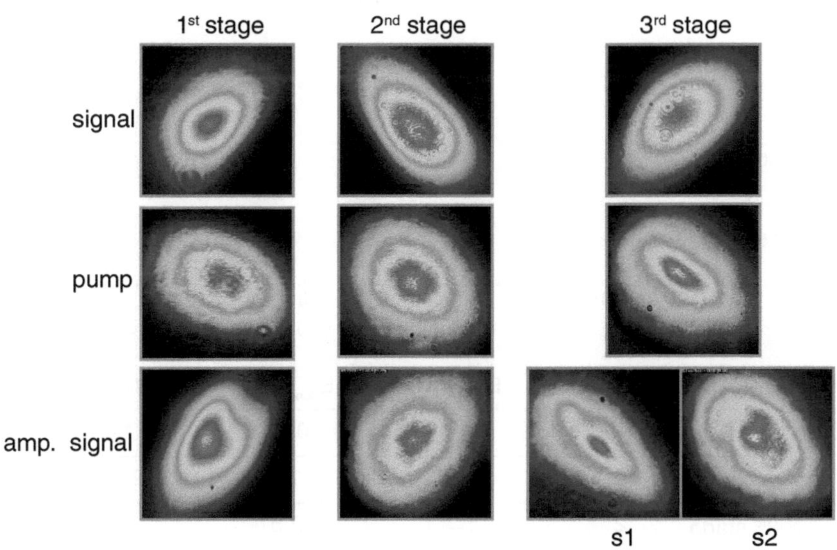

Fig. 5.3 Beam profiles of the signal and pump in the OPCPA stages. The ellipticity of the signal beam is caused by the telescope. s1 shows the beam profile of the amplified signal when second stage of the OPCPA runs slightly below saturation. s2 shows the beam profile of the amplified signal when the second stage of the OPCPA runs in saturation

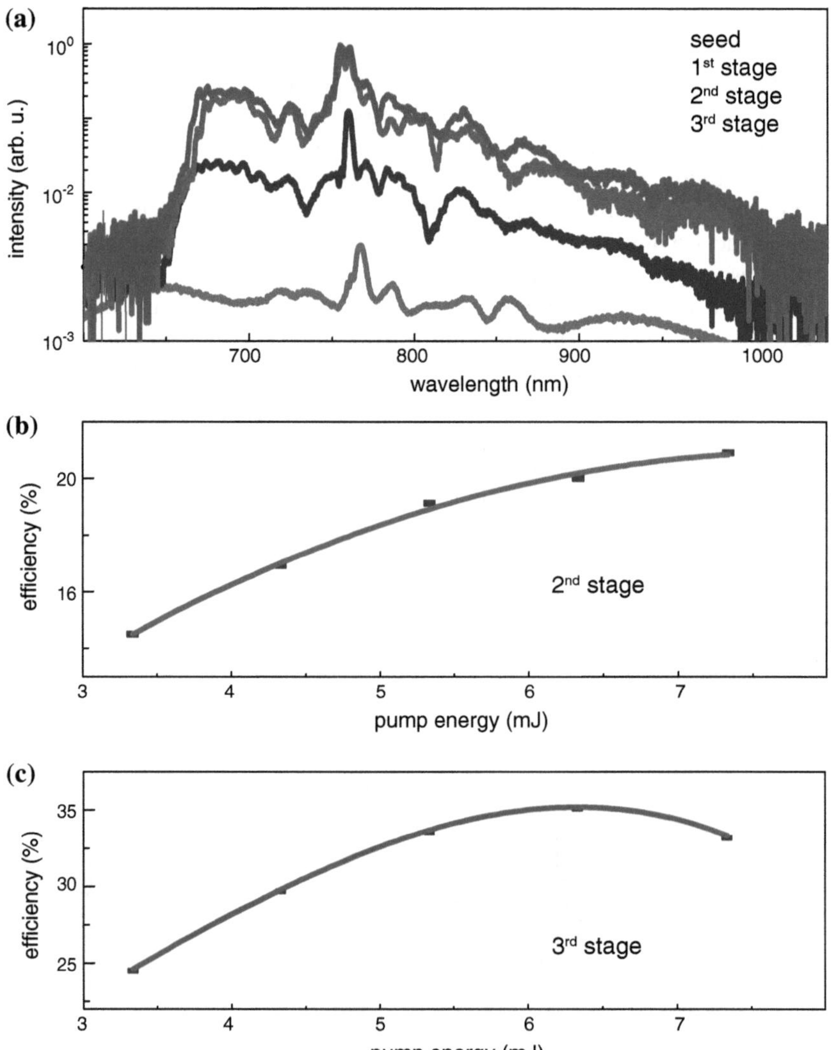

(d)

	L_c(mm)	E_{pump}(mJ)	E_{signal}(mJ)	efficiency
1st stage	1.5	1	0.12	12 %
2nd stage	2	7.3	1.7	20.5 %
3rd stage	2	–	2.5	30.1 %

Fig. 5.4 **a** The seed spectrum and the amplified spectra for three OPCPA stages, normalized to the output energy of each stage. **b** OPCPA efficiency curve at the second stage. **c** OPCPA efficiency curve at the third stage. To calculate the efficiency in (**b**) and (**c**) the net amplified signal energy is divided to the pump energy at the corresponding OPCPA stage. **d** The detailed parameters of each OPCPA stage. To calculate the efficiency of each stage, the net amplified signal energy is divided by the total pump energy of the OPCPA chain

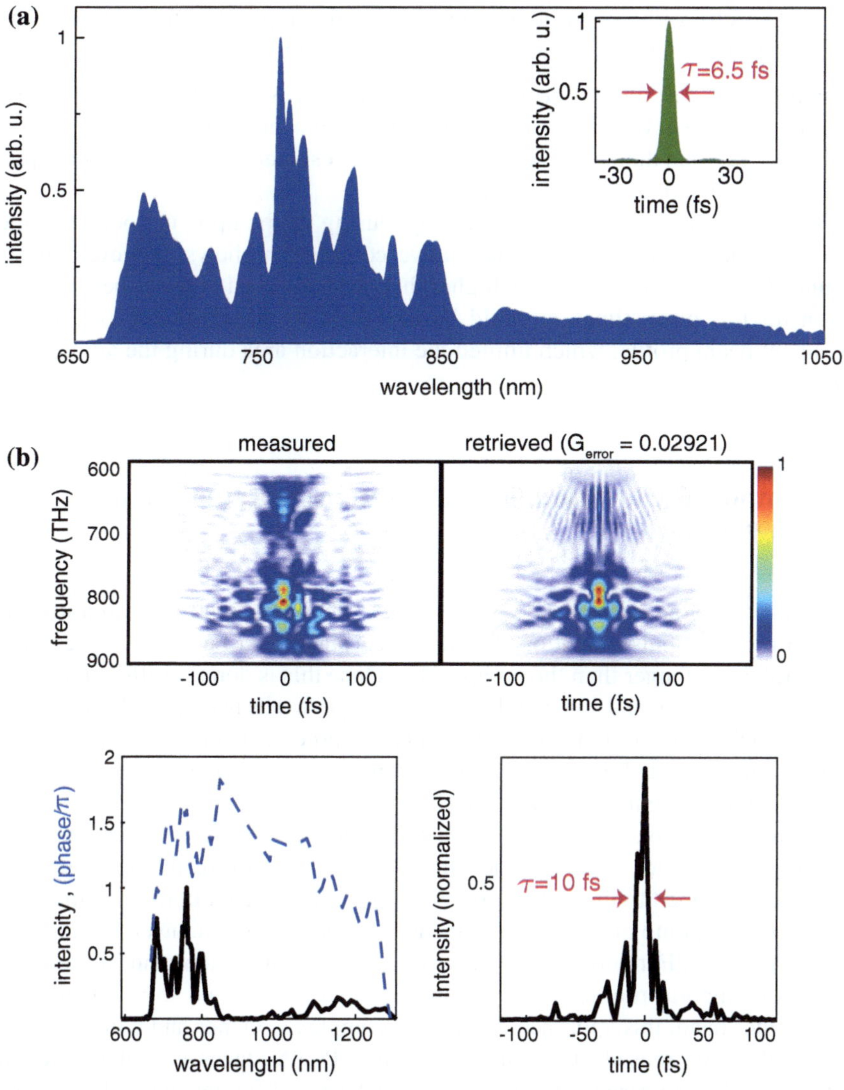

Fig. 5.5 **a** Amplified spectrum after 3 OPCPA stages on a linear scale, measured with a Si-based spectrometer. *Inset* The Fourier transform limit waveform corresponding to the measured spectrum. **c** SH-FROG measurement of the compressed pulse utilizing 12 reflections in a double angle chirped-mirror compressor. The retrieved spectrum shows the amplification of spectral components above 1 μm, not detectable with the available spectrometer. The pulse is compressed to 10 fs

duration of 10 fs. The FROG trace and its retrieved counterpart are shown in Fig. 5.5. It can be clearly seen from the retrieved phase and the presence of satellite pulses that the compression was not optimal. Higher-order dispersion and satellite pulses can be compensated by using specially designed chirped mirrors for this system. Also the retrieved spectrum shows amplification of spectral components above 1 μm, which were not measurable with the available Si-based spectrometer.

Conclusion By careful optimization of the pump intensity on the crystal, temporal stretching ratio between pump and seed, and spatial overlap between the pump and seed via large beam diameters, and by reusing the residual pump energy a high pump to signal conversion efficiency is achieved.

The calculated increase in efficiency by reusing the pump is in good agreement with the experimental results. Quantitative comparison shows, however, that the simulated conversion efficiency is higher than that observed in the experiment. The reason for this overestimation could be explained by the ellipticity of the pump and signal beam profile, which limited the interaction area during the amplification process.

5.2 Temporal Allocation of Pump Intensity Remainder (TAPIR)

In contrast[2] to conventional OPCPA systems, where the seed-pulse duration is a fraction of pump-pulse duration, OPCPA seed pulses can be linearly stretched to be several times longer than the pump-pulse. Once this is done, a different temporal delay between pump and signal pulses in each stage results in temporal and therefore spectral stacking of the signal in the amplification process. In this context, "stacking" refers to the controlled deposition of pump energy in subsequent temporal/spectral locations along the chirped seed pulse (Deng and Krausz 2013).

This technique enables shaping of the amplified spectrum to overcome the phase-matching deficiencies in broadband OPCPAs, as the phase-matching angle of the crystal can be tuned to the central wavelength of the selected portion of the signal. By reusing the pump after each amplification stage, the total conversion efficiency can be increased. Furthermore, by controlling the amplification gain in each stage, the spatio-temporal profile of the pump pulses can be shaped to a flat-top pulse. On the other hand, this approach demands a complex stretcher and compressor, the B-integral of the reused pump in subsequent OPCPA stages is higher, and most importantly, for the case of unsaturated OPCPA gain, different spectral parts of the amplified signal have varying energy stability which results in spectral beating.

Figure 5.6 shows simulation results for three different OPCPA designs. In the first design, 7 mJ of pulse energy, centered at 515 nm, with a duration of 1.6 ps at FWHM and beam diameter of 1 mm at FWHM is used to amplify a fourth-order super-Gaussian broadband spectrum centered at 850 nm, with a pulse duration of 1.6 ps and input energy of 120 μJ. Here, for both the pump and signal, the same pulse duration is considered and a 0.68 mm BBO crystal with a phase-matching angle of 24.7°, noncollinear angle of 2.7° and d_{eff} value of 2.3168 pm/V is used as

[2]This technique is based on (Deng and Krausz 2013).

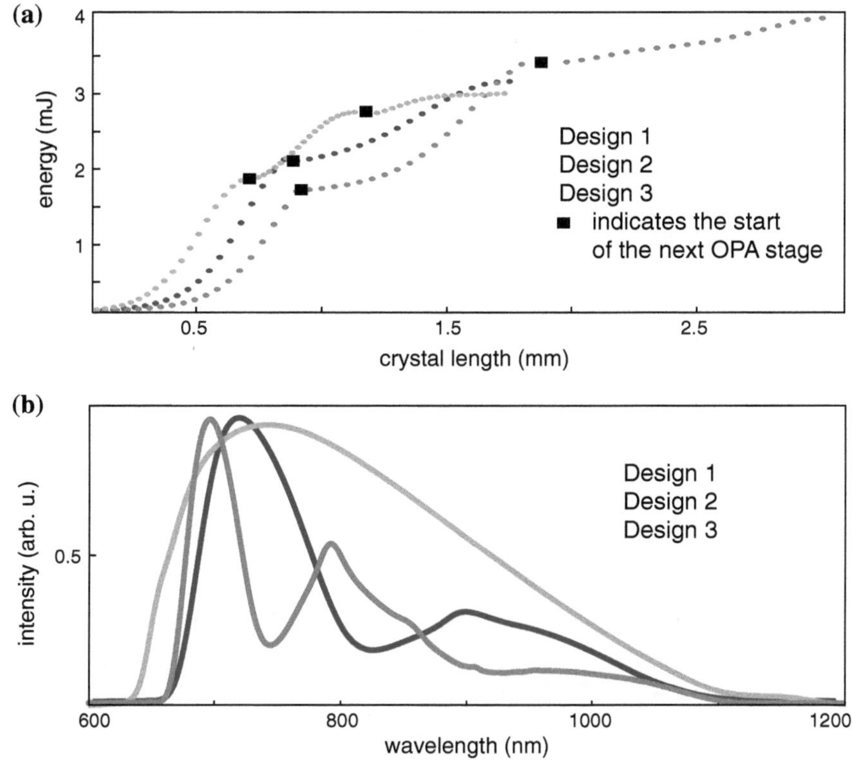

(c)

	Design 1			Design 2		Design 3		
	1st	2nd	3rd	1st	2nd	1st	2nd	3rd
efficiency (%)	24.6	22.4	11.8	28.5	26.8	22.6	35.2	30
total efficiency (%)	43			45		57		

Fig. 5.6 **a** The amplified pulse energy versus crystal length, **b** the amplified spectrum, **c** the conversion efficiency for three different OPCPA designs. In Design 1 the signal pulses are stretched to the same pulse duration as the pump pulses and amplified in 3 cascaded OPCPA stages, while the residual pump energy after each stage is used to pump the subsequent stage. In Design 2 the seed pulses are stretched to nearly twice the pump-pulse duration and amplified in two cascaded OPCPA stages and in Design 3, the signal-pulse duration is increased to more than three times the pump-pulse duration. In Designs 2 and 3 a different part of the signal spectrum is amplified in each OPCPA stage

the nonlinear medium. After the first amplification stage, the residual pump energy is used to pump the second stage consisting a 0.45 mm BBO crystal, while the pump beam diameter is resized and reduced to 78 % of its original diameter to keep the pump intensity at 100 GW/cm^2. Finally, the pump after the second stage is used to pump the 0.6 mm BBO crystal of the third stage while the beam diameter of both the pump and the signal are resized by a factor of 0.78.

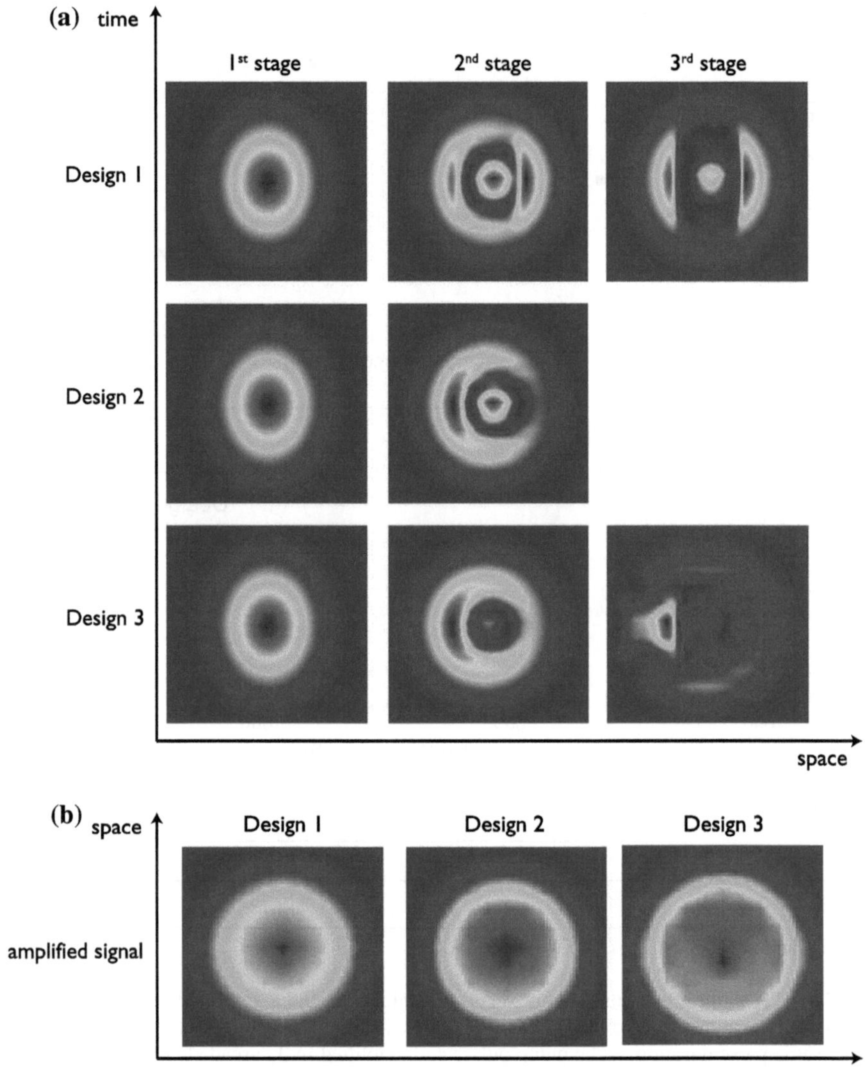

Fig. 5.7 **a** Spatio-temporal profile of the pump pulses for different OPCPA designs before each OPCPA amplification stage. **b** Beam profile of the amplified signal in each OPCPA design

Figure 5.7a shows the spatio-temporal profile of the pump before each amplification stage. Due to the Gaussian shape of the pump in time and space, the energy extraction took place primarily in the middle of the pump, leaving the wings mostly unaffected with a signature of energy backconversion in the center, owing to the fact that this part of the pump possesses the highest peak intensity. In this design, the OPCPA pump-to-signal conversion efficiency is increased to 43 % with a good spatio-temporal profile for the amplified signal (Fig. 5.7b). In the second design, the signal pulses were stretched linearly to 4 ps at FWHM and the blue part of the spectrum is amplified in a 0.85 mm BBO crystal by adjusting the temporal overlap between pump and signal pulses. Subsequently, the pump pulses after the first stage are reused to amplify the red part of the spectrum in a 1 mm BBO crystal at the second stage, while the pump beam size is reduced by a factor of 0.78. In this design the pump-to-signal conversion efficiency reaches 45 % and maintains the beam quality of the amplified signal (Fig. 5.7b).

The gain and the shape of the amplified spectrum can be further optimized by adjusting the phase-matching angles of the crystal in each stage to tune the amplification for the selected part of the spectrum (this option is not investigated in this study). The spatio-temporal profile of the pump before each OPCPA stage, shown in Fig. 5.7a, indicates better pump-energy-extraction but becomes asymmetric. The asymmetry is caused by the temporal walk-off between signal and pump pulses and can be suppressed by careful adjustment of the initial temporal overlap of the interacting beams. The 45 % conversion efficiency achievable from Design 2, is not drastically different from the achievable 38.6 % efficiency for the case where the signal and pump pulses have the same pulse duration (studied in Sect. 5.1) as the pump pulses after the first amplification stage in both cases maintain good spatio-temporal profiles.

As shown in Design 1, reusing the pump pulses for a third time results in a comparable efficiency. However, this changes for heavily stretched signal pulses. In design 3, the bluest part of the spectrum of the input signal pulses, which are linearly stretched to 9 ps pulse duration at FWHM, is amplified in a 0.9 mm BBO crystal. The residual pump diameter is reduced by a factor of 0.78 and used to amplify the middle part of the input spectrum in a 1 mm BBO crystal and finally, the red wing of the spectrum is amplified in a 1.2 mm BBO crystal while the input pump beam is shrunk by a factor of 0.68 before the last OPCPA stage. Here the pump-to-signal conversion efficiency reaches 57 %, which is a noticably higher value compared to the other designs. The spectral narrowing for Designs 2 and 3 visible in Fig. 5.6b, is caused by a suboptimal stretching factor of the input signal and phase-matching angle of the crystal. Optimizing these parameters are cumbersome in simulation but straightforward in an experimental setup.

Experiment An 8 mm-thick SF_{57} plate was used for stretching the amplified signal pulses after the first OPCPA stage, as described in Sect. 5.1. The glass was used at the Brewster's angle to minimize reflections from the surface of the glass. The delay between pump and signal, and the phase-matching angles of the BBO crystal at the second stage were adjusted to shift the amplification to the blue wing of the input spectrum and 4 W of power was obtained (blue curve in Fig. 5.8a). In the third OPCPA stage, the phase-matching and temporal overlap between pump and signal pulses were adjusted, to shift the gain to the red wing of the input spectrum with a total amplification power of 6 W (Fig. 5.8a, pink curve).

The same setup, after removing the bulk-stretcher, resulted in 6.2 W amplification after the third stage. Figure 5.8b compares the amplified spectra for these two cases.

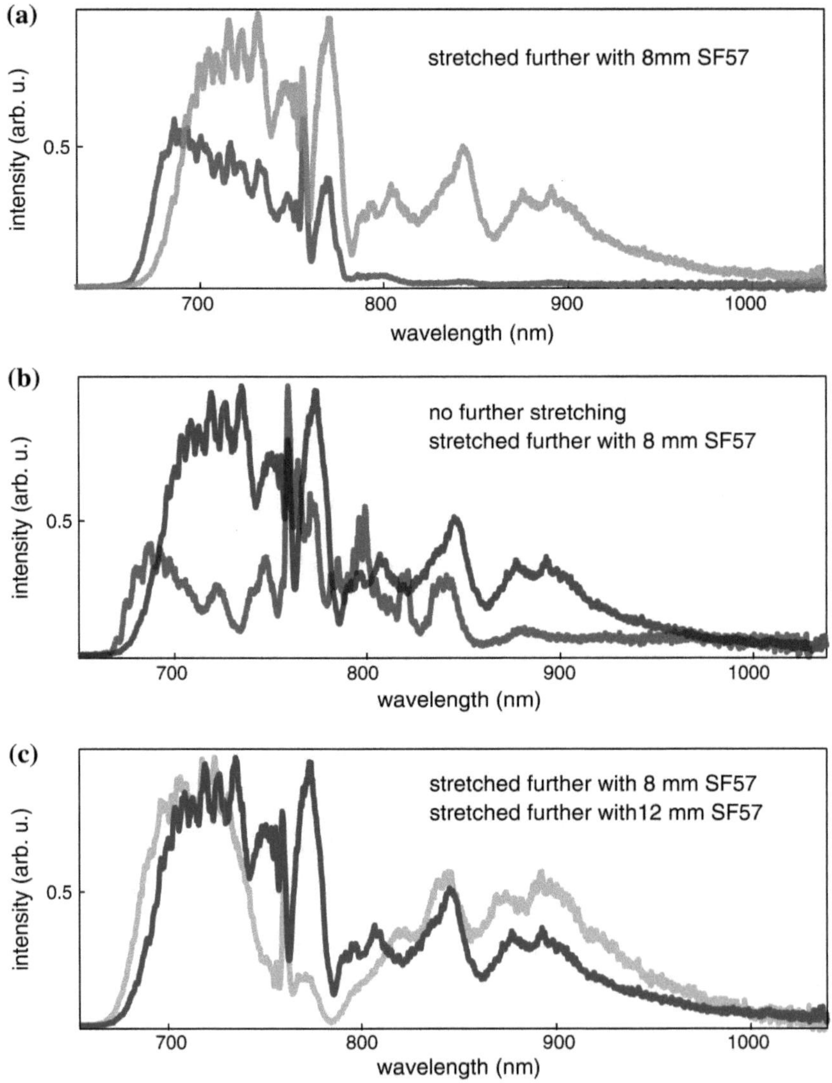

Fig. 5.8 **a** The stacked spectrum in a two-stage OPCPA system, temporally stretched by 8 mm of SF_{57}. The high frequencies were amplified in the second stage (*blue curve*) and the low frequencies were amplified in the third OPCPA stage (*pink curve*), recycling the pump energy after the second stage. The amplified signal energy in this case is comparable to the amplified energy obtained from the OPCPA system without additional stretching. **b** Comparison between the amplified spectra in a stacked OPCPA (*blue curve*) and with no additional stretching and reusing the pump (*gray curve*). **c** The amplified spectrum of a pulse stretched using 8 mm (*blue curve*) and 12 mm SF_{57} (*purple curve*) and stacked in two-cascaded OPCPA stages

The amplified spectrum of the stacked-OPCPA is smoother and has a square shape as the stretching of the signal spectrum provides comparable amplification gain for different frequencies. The red wing of the spectrum gains more energy in this case, as the phase-mismatching could be minimized (see Fig. 2.5). Also the spike around 780 nm is heavily suppressed, as further stretching of the signal reduces the temporal overlap between different frequencies. Further stretching of the signal in a 12 mm-thick SF_{57} resulted in a reduced amplified power of 2.75 W in the second stage, 5 W in the third stage, and produced a hole in the amplified spectrum at 790 nm (Fig. 5.8c), leaving the temporal stretching of the seed using 8 mm-thick SF_{57}, the optimum.

Conclusion Temporal stretching of the signal pulses to twice the pump-pulse duration, demonstrated in this section experimentally, did not show any increase in the OPCPA conversion efficiency compared to the scheme realized in Sect. 5.1. The simulated conversion efficiency for these two schemes is in the same range due to the fact that the residual pump energy after only one amplification stage (without any extra temporal stretching of the signal pulses) preserves good spatio-temporal quality (Fig. 5.7).

As the difference between simulated amplified energy in these two cases is only on the order of 1 W of average power, their experimental evaluation demands precise control over the signal-pulse duration.

However, stretching the duration of the signal pulse to three times (or more) the pump-pulse duration results in a significant increase in conversion efficiency (Fig. 5.6). Here, the advantage of this technique becomes distinguishable, as the residual pump pulses after the second amplification stage possess a better temporal and spatial quality for the heavily chirped signal pulses in comparison to the case discussed in Sect. 5.1. The experimental demonstration of the latter case could be an interesting study for the next generation of Ph.D. students!

5.3 Bandwidth Engineering

To optimize the amplification bandwidth in an OPCPA system, different crystals or a crystal with different phase-matching angles can be utilized. Figure 5.9a compares the simulated amplified spectrum from a 2.75 mm-thick LBO crystal at a phase-matching angle of 14.4° and the noncollinear angle of 1.1° with an amplified spectrum from a 1 mm-thick BBO crystal with a phase-matching angle of 23.3° and a noncollinear angle of 0°, and a 1 mm-thick BBO crystal with a phase-matching angle of 24.5° and a noncollinear angle of 2.7°.

In this study, the pulse duration of the signal and pump pulses are both 1.6 ps at FWHM. The 3 mJ pulses with a beam diameter of 1.2 mm at FWHM are used as a pump and the input seed energy is 3 μJ. The same temporal stretching is used for the three cases. Combining the two phase-matching angles in a BBO crystal results in broader amplification compared to LBO. Figure 5.9b studies the amplification gain in a cascaded OPCPA stage using these two different phase-matching angles. Firstly,

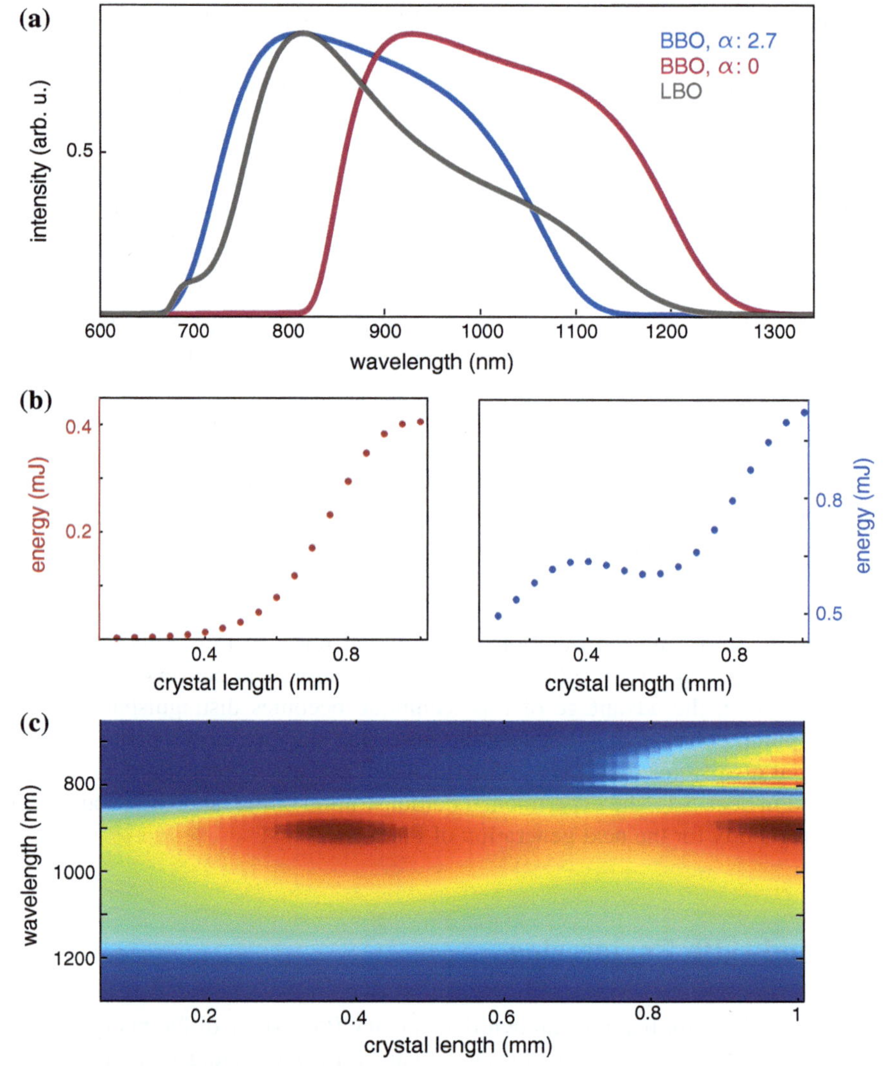

Fig. 5.9 **a** Simulated amplified spectra in different crystals and different phase-matching and noncollinear angles. **b** The OPCPA gain in a 2-stage OPCPA, using two BBO crystals with two different sets of phase-matching angles. *Left* the amplified energy versus crystal length in a BBO crystal with a noncollinear angle of $\alpha = 0°$ in the first stage. *Right* The amplified energy versus crystal length in a BBO crystal with a noncollinear angle of $\alpha = 2.5°$ in the second stage. **c** The amplified spectrum versus crystal length in the second stage

the red wing of the spectrum is amplified in the first stage at a noncollinear angle of 0°, and the blue wing of the spectrum in the second stage where the noncollinear angle is set to 2.7° and pumped with 3 mJ of energy.

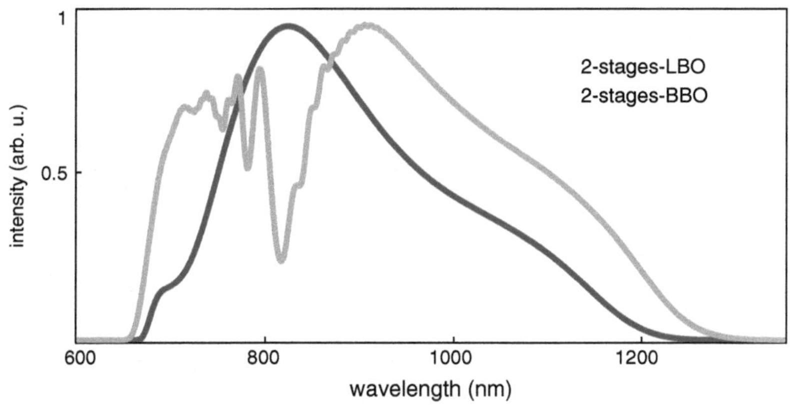

Fig. 5.10 Amplified spectra in a 2-stage OPCPA using LBO crystals (*blue curve*) and BBO crystals with different phase-matching angles (*purple curve*)

The amplification behavior in the second stage is very interesting. Initially, the spectral components that were amplified in the first stage and are phase-matched in the second stage, namely 800–1100 nm, are amplified until they reach saturation. Further amplification of these spectral components is due to the reset of the idler in the second stage. In broadband OPCPA, energy saturation of each frequency component rarely happens at the same crystal position. Later, energy backconversion takes place and at a crystal thickness of 0.6 mm, the amplification is shifted to the blue wing of the input spectrum (Fig. 5.9b, c). Figure 5.10 compares the amplified spectra in the two-stage OPCPA using LBO crystal or using BBO crystals at different phase-matching angles. The amplified energy in both cases is similar. 1.1 mJ in the LBO crystals and 1 mJ in the BBO crystals for a total pump energy of 6 mJ. The amplified spectrum for the latter case is broader, which reduces the Fourier transform limit of the amplified signal from 5.3 to 4.4 fs.

Experiment The broadband, octave-spanning spectrum described in Sect. 3.2 is amplified in an OPCPA system similar to the system described in Sect. 5.1. A 2 mm LBO, 2 mm BBO, and 3 mm LBO crystal was used in each amplification stage alternatively. The first stage was optimized to amplify a broad spectral range spanning from 750 to 1400 nm up to 50 μJ of energy. In the second stage, the measured amplified spectrum using a Si-based spectrometer, spans from 670 to 1100 nm and contained 1.1 mJ of energy. Finally at the third stage, the amplified energy was scaled to 1.8 mJ using the LBO crystal. The spectra shown in Fig. 5.11 are normalized to the energy in each amplification stage. Comparing the spectra after the second and third stages indicates energy backconversion for spectral components that reached saturation in the previous stage. The system has a lower efficiency compared to the system described in Sect. 5.1, as the main goal in this study was the engineering of the spectral bandwidth and the parameters of the system are not optimized in terms of the conversion efficiency.

(a)

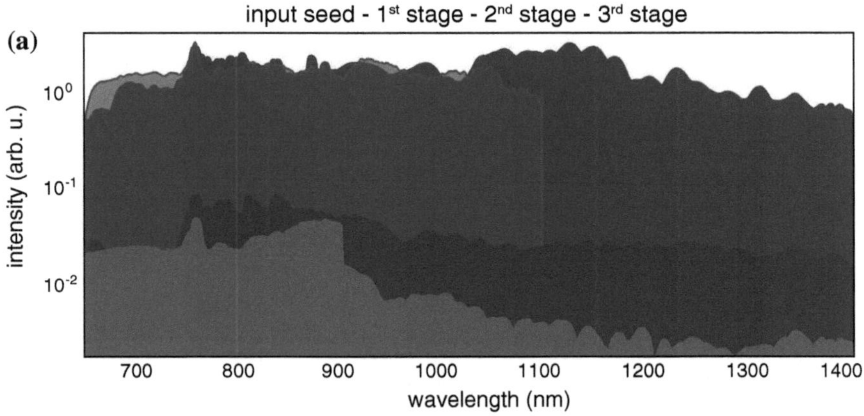

(b)

	L_c(mm)	E_{pump}(mJ)	E_{signal}(mJ)	efficiency
1st stage	2/ LBO	1	0.05	5 %
2nd stage	2/ BBO	7	1.1	16 %
3rd stage	3/ LBO	–	1.8	22 %

Fig. 5.11 **a** Amplified spectra in a 3-stage OPCPA. The seed pulses in the first stage were amplified to 50 μJ using a 2 mm LBO crystal. In the second stage, the spectral components below 1100 nm were amplified in a 2 mm BBO crystal up to 1.1 mJ of energy, and finally in the last stage, the amplified energy was boosted to 1.8 mJ using a 3 mm LBO crystal. The OPCPA seed was generated via cascaded spectral broadening in a HCF and a 4 mm YAG crystal. **b** The parameters of the OPCPA system

The amplified spectrum supports a FWHM transform limited pulse duration of 4.3 fs. A set of double-angle chirped mirrors designed for spectral wavelengths of 700–1300 nm (Fig. 5.12a) was used for the preliminary compression of the amplified pulses. Figure 5.12b shows the SH-FROG measurement of the compressed OPCPA seed pulses without any amplification, using 12 reflections on the chirped mirrors. The seed pulses were compressed down to 8 fs. The red tail of the spectrum generated via filamentation in the YAG crystal could not be characterized as it contained very little energy compared to the blue part of the spectrum generated via SPM in a HCF. The pre-pulse visible in the SH-FROG trace and also the retrieved pulse in the time domain, indicates uncompensated third-order chirp. After amplification, the signal pulses were characterized after the second and third OPCPA stages. Figure 5.13a shows the pulse compression after the second OPCPA stage. The higher-order phase developed during the amplification can be seen in the measured FROG trace. The retrieved spectrum includes spectral components from 680–1100 nm as the energy in the red wing of the spectrum is much lower than the blue wing. Figure 5.13b shows the pulse compression down to 9 fs after the third amplification stage. The retrieved spectrum is in good agreement with the measured OPCPA spectrum shown in Fig. 5.14 (red curve). Pulse compression to the Fourier transform limit would

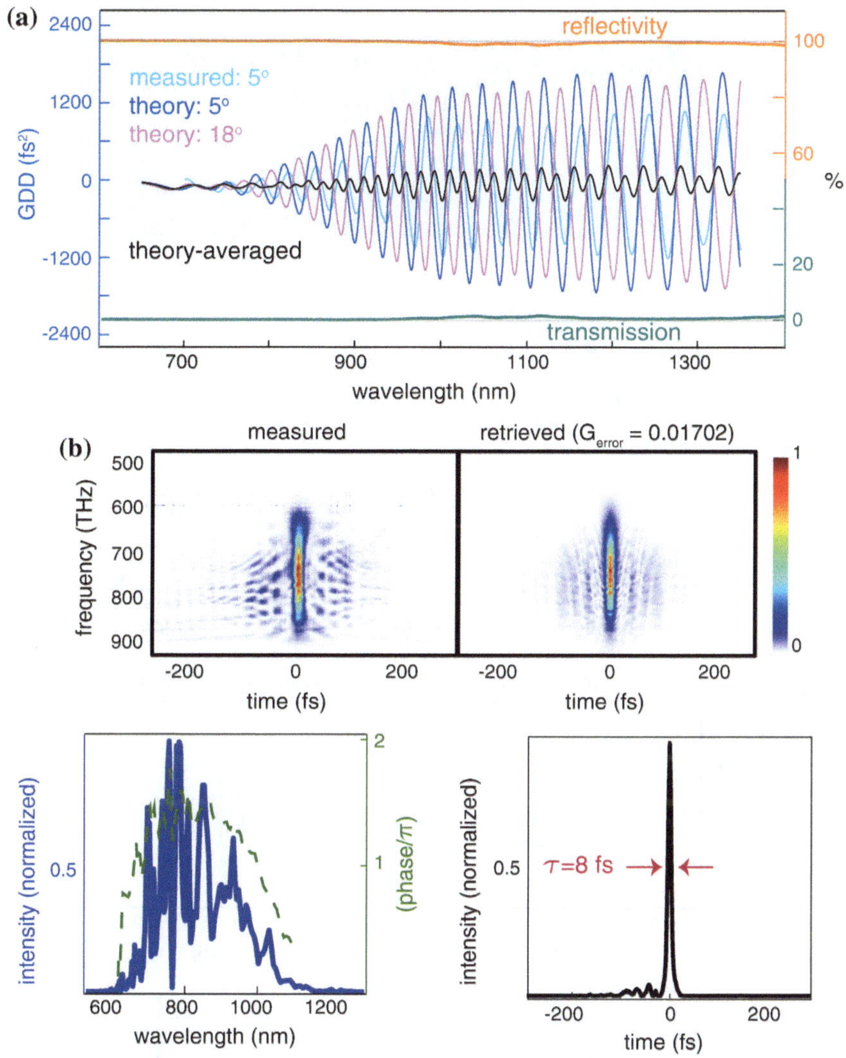

Fig. 5.12 **a** The GDD, reflectivity and transmission of the double angle chirped mirrors used for compression of the OPCPA system. The *black curve* shows the average GDD of the two complementary mirrors at two different angles of incidence. **b** The SH-FROG measurement of an OPCPA seed without any amplification, using 12 reflections on the chirped-mirror compressor

require an additional chirped-mirror compressor for compensating the higher-order OPCPA phase and a longer beam time!

Conclusion As shown in this section, utilization of well-selected, different nonlinear crystals extends the OPCPA-gain bandwidth substantially. The realized 3-stage OPCPA system, using one BBO and two LBO crystals, delivers 1.8 mJ pulses with a

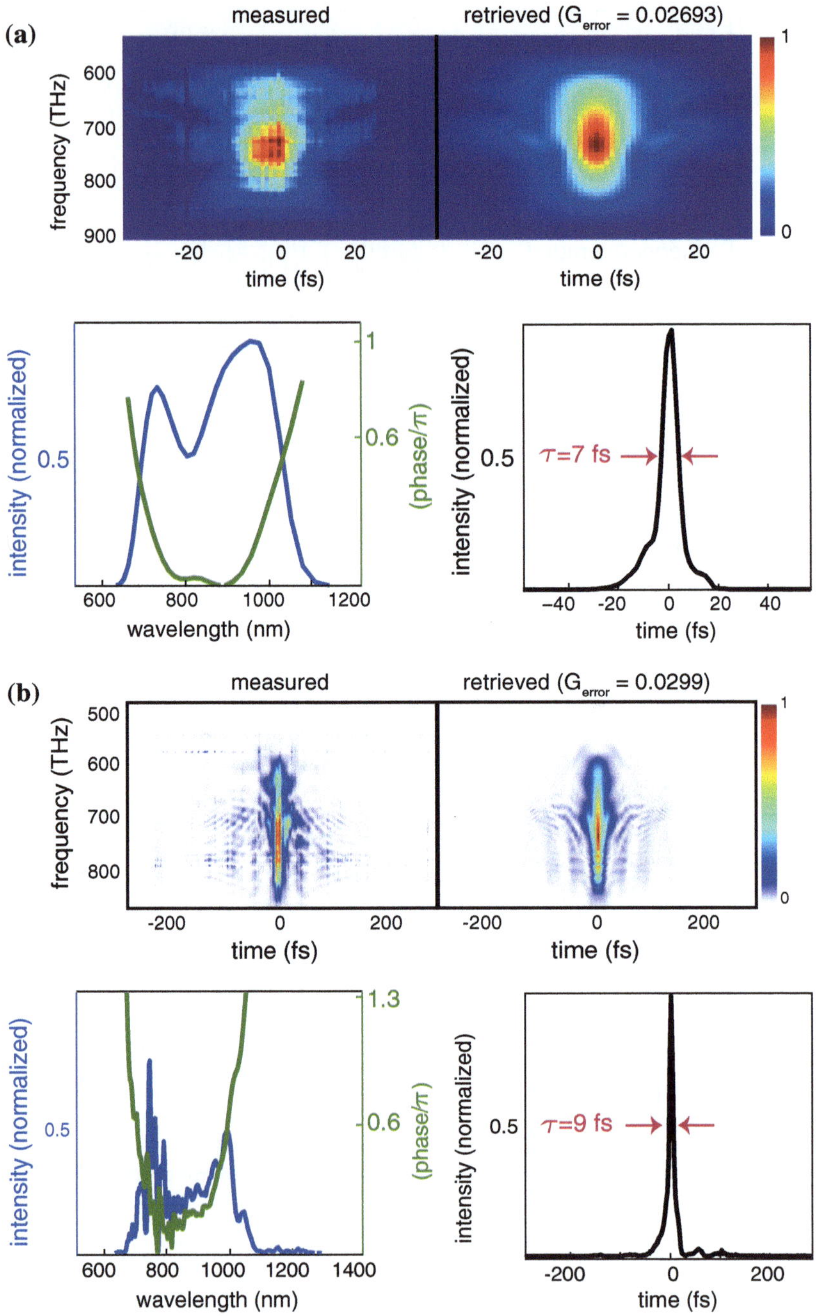

Fig. 5.13 Measured and retrieved SH-FROG traces and the retrieved spectrum and temporal profile of the pulses at the second (**a**) and third (**b**) stage of the OPCPA measured at an amplified energy of 1.1 mJ. The retrieved spectrum after the third stage is in a good agreement with the measured spectrum shown in Fig. 5.14

(a)

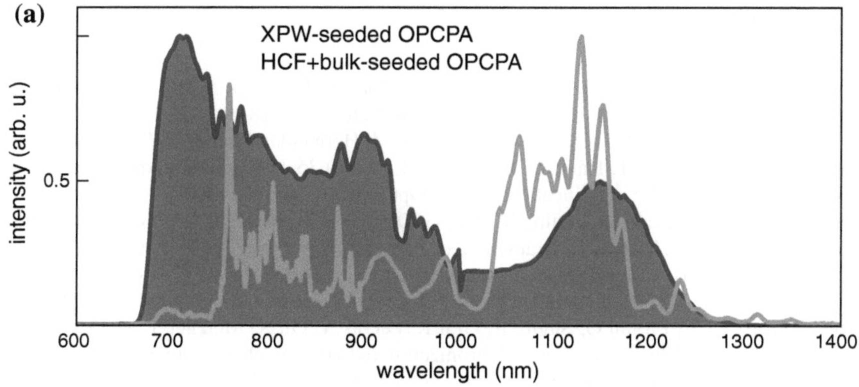

(b)	L_c(mm)	E_{pump}(mJ)	E_{signal}(mJ)	efficiency
1st stage	2	1.5	0.03	2 %
2nd stage	3	7	0.9	13 %
3rd stage	3	–	1.2	15 %

Fig. 5.14 **a** Amplified broadband signal generated in XPW process (*blue area*) and the amplified broadband signal generated in cascaded HCF-bulk stages (*red line*). **b** The parameters of the 3-stage OPCPA system seeded with a broadband signal generated via XPW

Fourier transform limit of 4.5 fs and thus supports even shorter pulse-duration than a 3-stage OPCPA system based on LBO alone, which supports only 5.3 fs. The reported extension of the supported amplified spectral bandwidth is crucial for experiments which rely on high-energy, single-cycle pulses.

The presented simulation results indicate that rather than using different crystal types, an equally impressive result can be achieved using BBO crystals at different phase-matching angles.

5.4 XPW Seeded OPCPA System

The generated broadband seed in Sect. 3.3 is amplified in the 3-stage OPCPA system described in Sect. 5.1 utilizing LBO crystals in all amplification stages. Figure 5.14 shows the amplified spectrum compared to the one described in Sect. 5.3. The smooth amplified spectrum does not contain sharp features, as the temporal contrast of the generated pulses in the XPW process is enhanced. Due to the lower input seed energy in this case compared to the previous OPCPA systems, the pump distribution and crystal thickness in each OPCPA stage should be optimized to achieve the highest conversion efficiency. This proof of principle study shows the potential of XPW as a temporal filter and a source for generating coherent photons at new frequencies, desired for seeding OPCPA systems.

References

Bagnoud V, Begishev IA, Guardalben MJ, Puth J, Zuegel JD (2005) 5 Hz, $>$ 250 mJ optical parametric chirped-pulse amplifier at 1053 nm. Opt Lett 30(14):1843–1845

Chekhlov OV, Collier JL, Ross IN, Bates PK, Notley M, Hernandez-Gomez C, Shaikh W, Danson CN, Neely D, Matousek P, Hancock S, Cardoso L (2006) 35 J broadband femtosecond optical parametric chirped pulse amplification system. Opt Lett 31(24):3665–3667

Deng Y, Schwarz A, Fattahi H, Ueffing M, Xun G, Ossiander M, Metzger T, Pervak V, Ishizuki H, Taira T, Kobayashi T, Marcus G, Krausz F, Kienberger R, Karpowicz N (2012) Carrier-envelope-phase-stable, 1.2 mJ, 1.5 cycle laser pulses at 2.1 μm. Opt lett 37(23):4973–4975

Deng Y, Krausz F (2013) Method and device for optical parametric chirped pulse amplification

Fattahi H, Teisset CY, Pronin O, Sugita A, Graf R, Pervak V, Gu X, Metzger T, Major Z, Krausz F, Apolonski A (2012b) Pump-seed synchronization for MHz repetition rate, high-power optical parametric chirped pulse amplification. Opt Express 20(9):9833–9840

Fattahi H, Skrobol C, Ueffing M, Deng Y, Schwarz A, Kida Y, Pervak V, Metzger T, Major Z, Krausz F (2012a) High efficiency, multi-mJ, sub 10 fs, optical parametric amplifier at 3 kHz. In: Conference on lasers and electro-optics 2012. OSA, Washington, D.C., CTh1N.6

Guardalben M, Keegan J, Waxer L, Bagnoud V, Begishev I, Puth J, Zuegel J (2003) Design of a highly stable, high-conversion-efficiency, optical parametric chirped-pulse amplification system with good beam quality. In: Optics Express 11.20, pp 2511-24

Ishii N, Turi L, Yakovlev VS, Fuji T, Krausz F, Baltuska A, Butkus R, Veitas G, Smilgevicius V, Danielius R, Piskarskas A (2005) Multimillijoule chirped parametric amplification of few-cycle pulses. Opt Lett 30(5):567–569

Schultze M, Binhammer T, Palmer G, Emons M, Lang T, Morgner U (2010) Multi-μJ, CEP-stabilized, two-cycle pulses from an OPCPA system with up to 500 kHz repetition rate. Opt Express 18(26):27291–27297

Schwarz A, Ueffing M, Deng Y, Xun G, Fattahi H, Metzger T, Ossiander M, Krausz F, Kienberger R (2012) Active stabilization for optically synchronized optical parametric chirped pulse amplification. Opt Express 20(5):5557–5565

Waxer LJ, Bagnoud V, Begishev IA, Guardalben MJ, Puth J, Zuegel JD (2003) High-conversion-efficiency optical parametric chirped-pulse amplification system using spatiotemporally shaped pump pulses. Opt Lett 28(14):1245–1247

Chapter 6
Design of a Multi-terawatt Field Synthesizer (LWS-pro)

Near[1] 1ps-pulse-pumped OPCPA offers several advantages over both long-pulse-driven OPCPA and conventional CPA implemented in solid-state laser amplifiers. As discussed in Sect. 2.1, in short-pulse-pumped OPCPA crystals can be pumped at much higher intensities, allowing high gains to be realized with very thin crystals, i.e. in combination with broad amplification bandwidths. Also, the short pump window greatly simplifies the implementation of CPA and dispersion control and, finally and most importantly, improves the temporal contrast of the amplified signal dramatically on the nanosecond to few-picosecond time scale.

The energy scalability of near-1-ps pulses of Yb:YAG thin-disk regenerative amplifiers, their near-diffraction-limited beam and excellent pulse energy stability, makes them promising pump sources for future OPCPA systems.

The gain bandwidth of an Yb:YAG-pumped OPCPA system can be extended up to several octaves, by using different crystals or crystals with different orientations yielding shifted gain bands (as it was experimentally and theoretically demonstrated in this chapter) or utilizing multiple pump beams at different wavelengths, i.e. at 1030 nm and its low-order harmonics at 515 and 343 nm (see: Sect. 2.2).

In what follows, the feasibility of super-octave, high-energy, high-power OPCPA sources and their expected performance when being pumped with several-kHz, kW-class thin-disk lasers is discussed.

Driving different OPCPA channels by 1030-nm picosecond pulses from Yb:YAG thin-disk amplifiers and their second and third harmonics, yield few-cycle pulses in the MIR, NIR and VIS spectral ranges, respectively (see: Sect. 2.2). These few-cycle pulses may also be superimposed on each other for the synthesis of multi-octave light transients. These options call for a seed coming in the form of a coherent, phase-stable, multi-octave supercontinuum covering the entire wavelength range of interest.

[1]This chapter is adapted from (Fattahi et al. 2014a).

H. Fattahi, *Third-Generation Femtosecond Technology*,
Springer Theses, DOI 10.1007/978-3-319-20025-5_6

Prototypical OPCPA architecture The common backbone for the proposed and numerically analyzed prototypical architecture in this chapter consists of (i) a high-power femtosecond laser, in our case a 100-W-scale Kerr lens mode-locked (KLM) Yb:YAG thin-disk oscillator, followed by (ii) the multi-octave seed generation, and (iii) a multi-100-W-to-kW-scale source of multi-mJ, near-1-ps-pulses, in our case based on Yb:YAG thin-disk amplifiers (Fig. 6.1).

Using chirped dichroic beam splitters, the supercontinuum seed is split into three spectral channels: VIS centered at 550 nm, NIR centered at 1 μm, and MIR centered at 2 μm. Each of the three OPCPA channels can be used to obtain few-cycle pulses. Their pump pulses are generated by a simple frequency-converter module comprising two LBO crystals. SHG in the first one yields twin pulses of comparable energy at 515 and 1030 nm, which are mixed in the second crystal to produce a third pump pulse at 343 nm by sum frequency generation (SFG). Moderate SHG and SFG conversion efficiencies ensure that all beams exiting the frequency converter unit have a good beam quality, which is important for OPCPA pumping (for more details, see: Sect. 6.1). The three beams are subsequently separated by dichroic beam splitters and directed into three OPCPA channels.

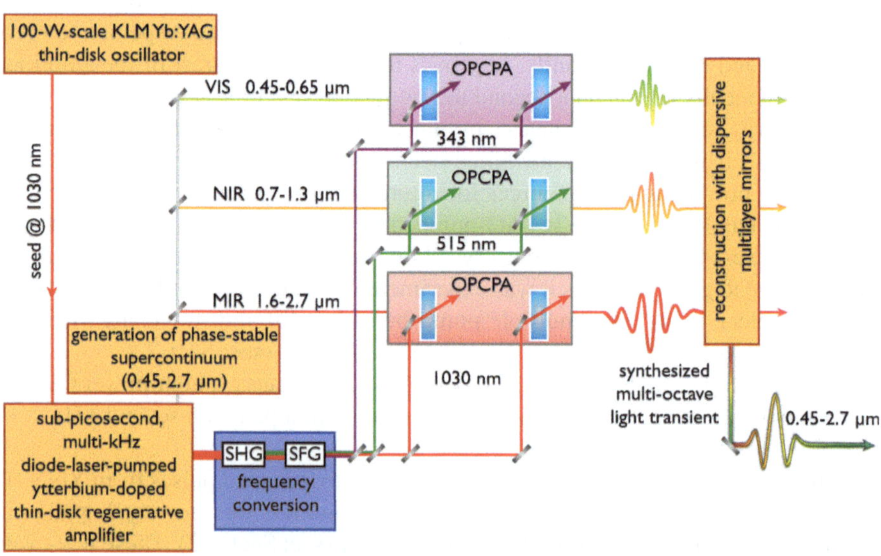

Fig. 6.1 Schematic architecture of a three-channel OPCPA system seeded and pumped by near-1-ps ytterbium lasers. A part of its output is used for generating the multi-octave supercontinuum signal, which is split into three channels, centered at 550 nm, 1 and 2 μm, respectively. The different channels are pumped by different low-order harmonics of the multi-mJ level, kHz, Yb:YAG regenerative amplifier output. Each channel supports few-cycle pulses. By coherently combining the three few-cycle channels, amplified, non-sinusoidal, multi-octave light transients can be generated

6.1 Proposed Multi-octave Seed Generation

DFG is the most powerful technique for the generation of MIR continua among the spectral broadening schemes discussed in detail in Sect. 3.4. This process creates a CEP-stable output from a non-CEP-stabilized pulse and has been successfully applied to seeding few-cycle MIR-OPCPA systems (Erny et al. 2007; Chalus et al. 2009; Deng et al. 2012; Hong et al. 2011). The DFG signal has good temporal contrast, is scalable in energy, and is passively CEP-stable which provides a reproducible shot-to-shot electric field, as required for HHG. The spectrum of the DFG output at mJ-level or μJ-level can be efficiently extended to cover the multi-octave range of 400–5000 nm in filament (Kartashov et al. 2012; Silva et al. 2012). Moreover, the spectrum of sub-20-fs pulses from a MIR OPCPA (Deng et al. 2012), could be efficiently broadened in a 1 bar-Ar-filled HCF, to cover the entire VIS and NIR spectral range from 400–2500 nm, as shown in Fig. 6.2. Such continua, at energy levels of hundreds of microjoules, could also be achieved with substantially longer, multi-cycle MIR fs pulses, in agreement with theoretical predictions (Kartashov et al. 2012). The temporal characterization of the generated continuum by SH-FROG shows the smooth spectral phase and its compressibility (Schwarz 2014).

In what follows, different designs for generation of broadband MIR, > μJ-level energy, passively CEP stabilized pulses, required for seeding the OPCPA are proposed.

1-ps-pumped Multi-octave Seed Generation The 10 fs pulses required to seed the continuum generator shown in Fig. 6.2, may possibly be derived directly from the output of the near-1-ps Yb:YAG pump source by cascaded temporal compression. This approach would eliminate the need for the active pump-seed synchronization system for the OPCPA, because both the pump and the seed would travel comparable optical paths. In a cascaded gas-filled-HCF, the 1 ps input pulses centered at 1030 nm could be broadened to 700–1500 nm (Voronin et al. 2013). Although in this scheme the spectral phase in the vicinity of 1030 nm is not well-behaved and could cause a long pedestal in the time domain, the spectral phase in the blue and red wings of the spectrum are well-behaved and suitable for a type II DFG process.

In an alternative approach, shown in Fig. 6.3a, 1-ps-pump pulses could be temporally shortened in an XPW stage (Cotel et al. 2006). As XPW acts as a gate in the time domain, no additional step would be required to compress the generated signal.

Recent experiments by Karpowicz et al. show that the red shift of the laser radiation necessitated by energy conservation in optical rectification for generating terahertz transients leads to pulse shortening. In their experiment, 1.6 ps pulses at 1030 nm were compressed to 300 fs, after propagation in a 5 mm [110] GaP and a grating compressor. This scheme can be exploited as the second broadening stage after XPW to shorten pulses down to 100 fs, a suitable pulse duration for spectral broadening in a gas-filled HCF. After compression, the 10 fs pulses are a suitable driver for the DFG process and multi-octave seed generation.

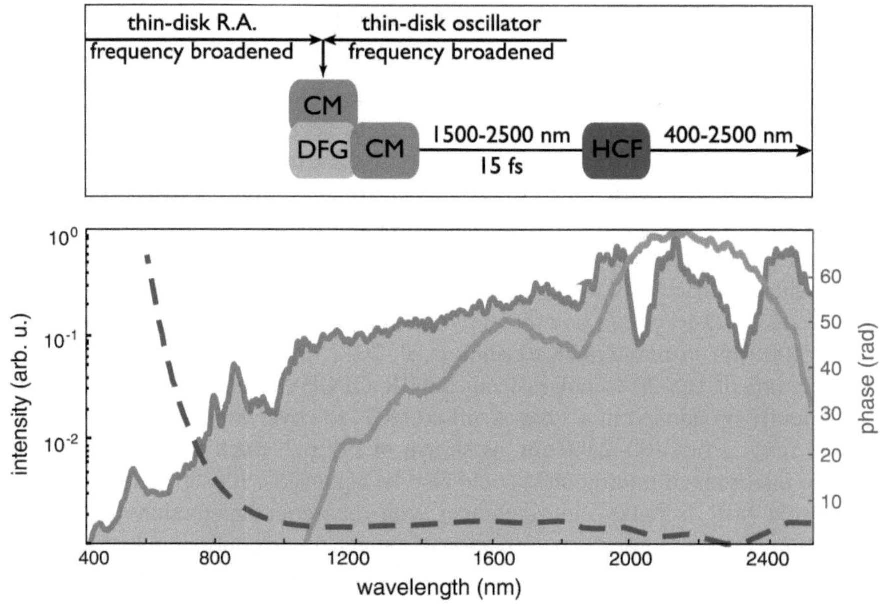

Fig. 6.2 Multi-octave seed generation. **a** Near-three-octave seed generation schemes based on the output from a KLM oscillator or driven by a ps thin-disk regenerative amplifier. The spectrally broadened and compressed output of both lasers drive the DFG stage. The resultant broadband DFG signal, centered at 2 μm is compressed and focused into a gas-filled HCF to extend the spectrum into the visible range. **b** Broadband phase-stable continua generated in preliminary experiments. Red curve: Difference-frequency radiation in a 500 μm-thick BBO crystal. Purple shaded area: Spectral broadening of a parametrically amplified three-cycle DFG signal in a gas-filled HCF to a continuum containing 33 μJ of energy (Schwarz 2014). The detected bandwidth of the spectrum in both cases is limited by the sensitivity of the spectrometer in the infrared tail. The dashed blue curve shows the measured spectral phase of the generated supercontinuum using the SH-FROG technique (Schwarz 2014)

Sub-200fs-pumped Multi-octave Seed Generation In this scheme, the 200 fs, μJ-level energy pulses directly from a KLM thin-disk oscillator, which serves to seed a thin-disk regenerative amplifier, are focused in a bulk (here YAG) for supercontinuum generation (Fig. 6.3b). The spectral components at 1030 nm and 650–750 nm are separated from the rest of the continuum. The latter one is amplified in an OPCPA stage pumped by the second harmonic of the thin-disk regenerative amplifier and then combined with the 1030 nm pulses in a DFG stage (Homann et al. 2012). As the efficiency of the generated supercontinuum in bulk does not exceed 2 %, the 1030 nm pulses have μJ-level energies and the energy of the DFG signal does not exceed several tens of nJ. Although the scheme is simple, it is not scalable to higher pulse energies.

In an alternative approach, the output of the oscillator is broadened in a cascaded SPM stages and compressed down to 10 fs pulses with μJ-scale energy (Fig. 6.3c) (Seidel et al. 2014). The spectrum of these pulses with a well-behaved spectral phase

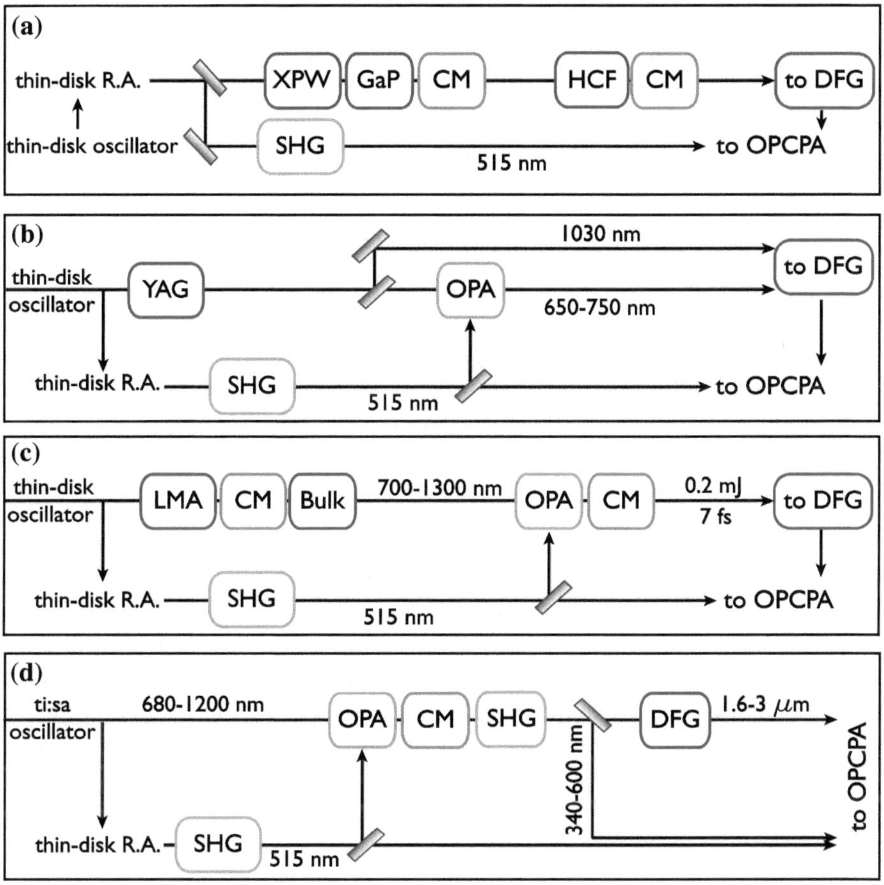

Fig. 6.3 **a** 1-ps-pumped broadband CEP-stable seed generation scheme based on Yb:YAG regenerative amplifiers and cascaded spectral broadening in an XPW stage, a GaP crystal and a HCF. **b** 200-fs pumped broadband CEP-stable seed generation scheme based on Yb:YAG oscillator and supercontinuum generation in YAG crystal. **c** Sub-200-fs pumped broadband CEP-stable seed generation scheme based on Yb:YAG oscillator and cascaded broadening in LMA fibers. **d** Ti:sapphire oscillator-based, multi-octave seed generation

and excellent spatial beam quality is matched to the gain band of BBO and LBO-based optical parametric amplifiers pumped by a fraction of the second harmonic of the Yb:YAG think-disk regenerative amplifier. Amplification in such an OPCPA to the several hundred μJ energy level and recompression of the amplified NIR pulse is followed by a DFG stage. Saturated amplification in such an OPCPA provides not only the pulse energy and intensity required for subsequent spectral broadening but also stabilizes the amplified NIR against energy fluctuations accumulating in the preceding nonlinear stages. Here the signal and pump pulses must be synchronized actively.

In the last proposed scheme (Fig. 6.3d), a broadband spectrum of a Ti:sapphire oscillator, used to seed the thin-disk regenerative amplifier, is amplified in an OPCPA stage. A broadband cascaded SHG and DFG of the amplified pulses after compression, extends the spectrum to VIS and MIR. In this scheme, active synchronization is required and the carrier-envelope phase of the VIS and MIR spectra need to be actively stabilized.

Conclusion By spectral broadening of broadband pulses centered in the mid-infrared and generated in a difference frequency process, a CEP-stable, multi-octave spectrum spanning from visible to mid-infrared with a high temporal contrast can be produced. This approach is the base of three different designs proposed in this section.

Deriving the required spectrum to pump a DFG stage from a 1-ps thin-disk regenerative amplifier, eliminates the need for a pump-to-signal active temporal synchronisation in the OPCPA chain. This spectrum can be achieved in a setup consisting of three nonlinear stages. However, the proposed scheme consists of three unsaturated nonlinear steps which makes achieving the required energy and spectral-bandwidth stability of the generated signal challenging.

The two other proposed approaches in this section, based on sub-300-fs pulses of a thin-disk oscillator require active temporal synchronisation of the signal and pump pulses. Here, the setup contains a saturated OPCPA stage which may provide necessitated energy and spectral bandwidth stability for the subsequent DFG stage.

For generality a multi-octave seed generation scheme based on a Ti:sapphire oscillator is discussed, as well. Here the generated spectrum in the mid-infrared is passively CEP-stable while the near-infrared and visible part of the spectrum has to be actively stabilized. In this scheme, an active temporal synchronization between pump and seed pulses is also required.

6.2 Synthesizer Channels

Each of the three prototypical, multi-terawatt OPCPA channels can be designed to generate few-cycle pulses at a single carrier wavelength, by amplifying and recompressing the continua seeded into the amplifier chains. The following model calculations for a prototypical multi-color, multi-terawatt, few-cycle system is performed by assuming the availability of a 200 mJ Yb:YAG thin-disk regenerative amplifier and designed for operating at a 5-kHz repetition rate (Fattahi et al. 2014a).

In this design study, the pump energy is distributed preferentially to the NIR and MIR channels. This strategy is motivated by numerous applications benefiting from longer wavelengths, like generating keV photon energies in HHG due to the λ^2-scaling of the high-harmonic cutoff energy, as illustrated in Sect. 6.2 (Tate et al. 2007).

Generation of Different Harmonics of the Pump Laser In a first step, SHG of the 1030 nm Yb:YAG laser pulses in a type I LBO crystal cut at 40.2° and the d_{eff} of 0.7 pm/V, is implemented with an optical-to-optical efficiency of approximately

50 % to share the pump-pulse energy equally between the fundamental laser wavelength and its second harmonic, i.e. 100 mJ of energy at 515 nm and 1030 nm each. LBO is chosen instead of BBO, owing to its availability in large sizes and its small spatial and temporal walk-off, in spite of its smaller effective nonlinear coefficient.

The direct THG in solids has low conversion efficiency reaching ≈6 % in BBO crystal (Banks et al. 2002), caused by the small magnitude of the $\chi^{(3)}$ nonlinear polarization compared to the one arising from $\chi^{(2)}$. Hence, for generating the third harmonic of the laser, cascaded third harmonic generation in a simple collinear geometry without the need of an additional delay element is used (Herrmann et al. 2010a). The SHG stage is followed by SFG in another LBO crystal with a conversion efficiency of approximately 20 %. This approach yields approximately 40 mJ at 343 nm, 74 mJ at 515 nm and 86 mJ at 1030 nm for pumping the VIS, NIR, and MIR channels of the OPCPA system, respectively (Fig. 6.4).

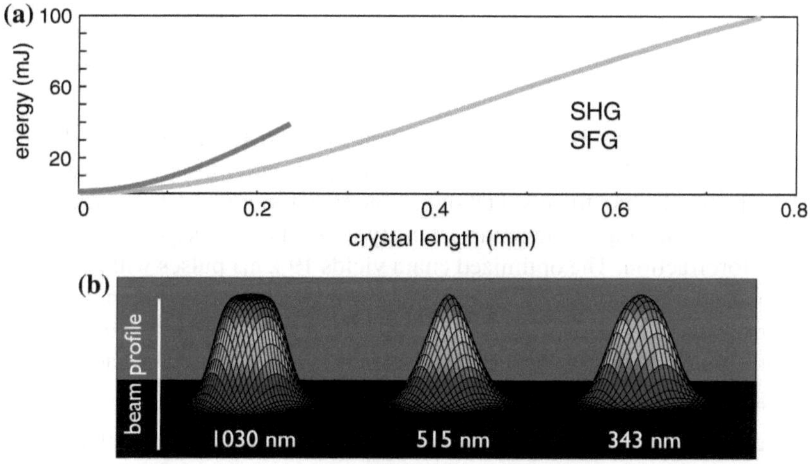

(c)

	d_{eff} (pm/V)	L_c(mm)	Φ(mm)	E_{input}(mJ)	E_{output}(mJ)	efficiency
SHG	0.82	0.75	9.8	200	102	50 %
					98	
SFG	0.7	0.23	D_{amp}×0.7	100	90	35 %
			D_{amp}×0.85	98	73	
					36	

*D_{amp}: diameter of the amplified beam

Fig. 6.4 **a** The SHG energy over crystal length for a near-1-ps, 200 mJ, 1030 nm Yb:YAG regenerative amplifier (*green curve*) and the SFG between the second harmonic and the fundamental (*blue curve*). **b** The simulated beam profile of the cascaded SHG-SFG frequency converter. **c** The detailed parameters of the simulation. L_c: length of the crystal, ϕ : beam diameter at FWHM, E_{input}: input energy, E_{output}: output energy. In the table, red areas indicate the parameters related to fundamental, while *green* and *blue areas* are related to the parameters of second and third harmonics of the laser, respectively

The OPCPA is designed for maximum spectral bandwidth and pump-to-signal efficiency. The latter has been optimized, among others, by recycling the residual of the pump energy after the penultimate amplifier stage, a scheme which was successfully implemented in an experiment discussed in detail in Sect. 5.1. In the last crystal of the chain, the phase-matching angle, noncollinear pump-signal angle, spatial walk-off and spatial overlap of the interacting beams is slightly readjusted for maximum efficiency. In the following simulations, a Gaussian spatial and temporal profile of the pump pulses is assumed.

A linearly-chirped fourth-order super-Gaussian spectrum is used for the seed pulses. The pump and seed duration in each channel is set to 1.6 ps at FWHM. After setting the pump-beam diameter such that the desired on-axis peak intensity is realized, the size of the seed beam is optimized for best efficiency and amplified beam profile.

In the spectrum shown in Fig. 6.2, the different spectral regions, 1600–2800 nm (MIR channel), 700–1400 nm (NIR channel) and 450–600 nm (VIS channel) contain 40, and 3.5 % of the total energy of the super-continuum, respectively. In this study, the same spectral energy distribution for the input seed containing a total energy of 6.5 μJ is assumed.

MIR Channel $LiNbO_3$ is chosen as the gain medium for a 4-stage MIR OPCPA chain (Sect. 2.2), with the first three stages pumped with pulses of energy of 1, 7 , 78 mJ, respectively, and the residual pump energy surviving the third stage recycled to pump the fourth stage. The pump intensity in all stages is kept at 70 GW/cm^2 to avoid photorefraction. The optimized chain yields 19.2 mJ pulses with a bandwidth-limited pulse duration of 12.3 fs at FWHM, corresponding to less than two cycles of the carrier wave (Table 6.1). A d_{eff} value of 3.3 pm/V is assumed in the simulation. The pump-to-signal energy conversion efficiency is 18 %, limited by back conversion of the signal into the pump via phase-matched SHG (Fig. 6.5).

NIR Channel In the second-harmonic-pumped NIR channel, a broad phase-matched bandwidth can be obtained by using a BBO crystal with an internal pump-signal angle of 2.2° and a phase-matching angle of 22°, or an LBO crystal with an internal pump-signal angle of 1.1° and a phase-matching angle of $\phi = 15.5°$ in the noncollinear type I geometry. LBO is chosen as the gain medium because it supports somewhat shorter amplified pulses. The NIR OPCPA chain also consists of 4 crystals being pumped with a total energy of 74 mJ. The d_{eff} value of 0.82 pm/V is assumed in the simulation. Adjusting the peak pump intensity to 110 GW/cm^2, 22.7-mJ pulses with a bandwidth-limited pulse duration of 4.8 fs have been predicted by our simulation (Table 6.1), corresponding to a conversion efficiency of 30 %. Thanks to the noncollinear geometry, parasitic backconversion is strongly reduced, giving rise to an excellent efficiency (Fig. 6.5).

VIS Channel Finally, the VIS channel pumped with 40 mJ, 343 nm pulses, has been modelled by using type-I BBO crystals phase matched at 32.7° and an internal noncollinear pump-signal angle of 4.5° in three amplifier stages. The pump intensity in all stages is kept to 70 GW/cm^2 to avoid BBO degradation. With pump energies

Table 6.1 Detailed parameters of the simulated channels

	MIR channel			
	1st stage	2nd stage	3rd stage	4th stage
E_p(mJ)	1	7	78	54
L_c(mm)	2.5	1.7	1.4	1.1
E_{s-amp}(mJ)	0.146	1.3	12	19.2
Φ_p(mm)	0.9	2.4	7.6	6.1
Φ_s(mm)	0.9	*$D_{amp}\times 3$	*$D_{amp}\times 3.5$	*$D_{amp}\times 1.2$
	NIR channel			
	1st stage	2nd stage	3rd stage	4th stage
E_p(mJ)	1	7	66	54
L_c(mm)	2.5	1.5	1.5	1.2
E_{s-amp}(mJ)	0.189	1.6	14	22.5
Φ_p(mm)	0.7	1.8	5.6	4.5
Φ_s(mm)	0.7	*$D_{amp}\times 2.5$	*$D_{amp}\times 3.2$	*$D_{amp}\times 1.2$
	VIS channel			
	1st stage	2nd stage	3rd stage	
E_p(mJ)	3	37	29.4	
L_c(mm)	1.2	0.75	0.5	
E_{s-amp}(mJ)	0.474	4.8	7.1	
Φ_p(mm)	1.2	5.3	4.66	
Φ_s(mm)	1.2	*$D_{amp}\times 4.4$	*$D_{amp}\times 1.1$	

E_p: input pump energy. L_c: crystal length. E_{s-amp}: amplified energy of the signal. Φ_p: beam diameter of the pump at FWHM. Φ_s: beam diameter of the signal at FWHM. D_{amp}: Diameter of the amplified beam at FWHM

of 3 mJ and 37 mJ for the first two OPCPA stages and reusing the residual pump energy of the second stage in the third stage, 7.1 mJ pulses with a bandwidth-limited duration of 5.1 fs could be reached (Fig. 6.5). A d_{eff} value of 1.56 pm/V is assumed for the calculation (Table 6.1, Fig. 6.6).

Challenges Although parametric amplification, due to the (ideally) instantaneous, non-resonant nature of the light-matter interaction involved should be scalable in average power, limitations and challenges might arise in real crystals due to the absorption of certain frequencies, even with a very small absorption coefficient, in high power systems. This could cause a temperature gradient in the crystal and eventually lead to cracking of the crystal or cause phase-matching changes via the thermo-optic refractive index of the crystal.

In LBO, similar to BBO, a problematic absorption may be expected from the second harmonic of the pump centered at 515 nm due to the high power density. However, this value is estimated to be negligible for the pump lasers with the average power as high as 1 kW and should not be problematic in this case (Rothhardt et al. 2013; Prandolini et al. 2014). In the MIR channel the linear absorption in $LiNbO_3$ at

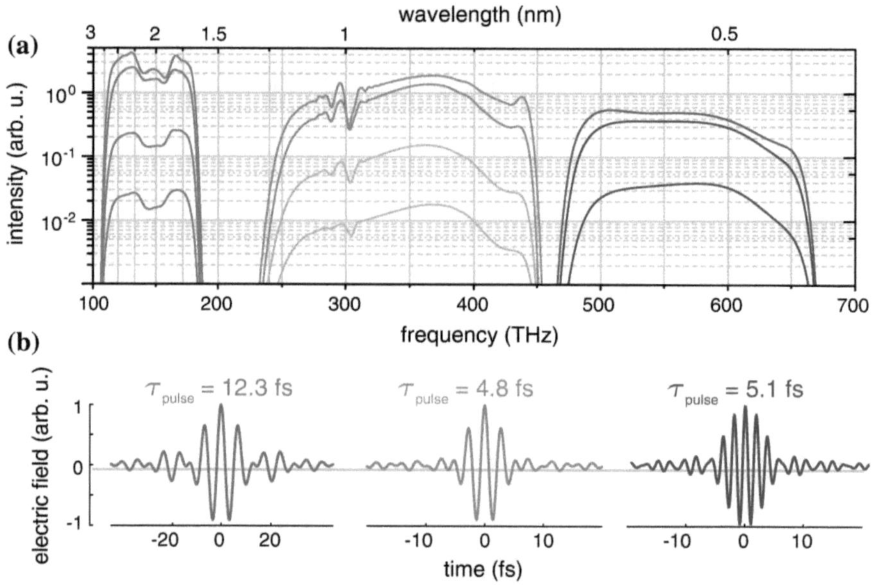

Fig. 6.5 **a** Amplified spectra of the three channels of the synthesizer obtained from simulation. The spectra of the different stages in the different channels are normalized to their energy and shown on a logarithmic scale. VIS channel is plotted in *blue*, NIR in *red-orange*, and MIR in *brown*. The dip in the amplified spectra of the MIR channel is caused by the phase-matched SHG of the signal in the degenerate OPCPA. The strong modulations around 1 μm in the NIR channel are also caused by SHG. Here, because of the noncollinear geometry, the dip is narrower. **b** Fourier-transform-limited electric field corresponding to the final output spectrum of each channel and their pulse duration at FWHM. All channels support few-cycle pulses

2.6 μm could cause a temperature gradient and eventually mechanically damage the crystal. This problem could be avoided by cutting the amplified spectrum to below 2.6 μm or heating up the crystal. The main challenge of the visible channel is caused by nonlinear absorption of the pump laser as its second harmonic approaches the UV cut-off edge of the BBO at 189 nm. However, the calculated two-photon absorption in the VIS channel of the synthesizer is negligible due to the low intensity employed. BBO crystal exhibits decreasing transmittance with increasing UV intensity and over time, e.g. when irradiated with ns-laser pulses centered at 262 nm, which could be improved by heating up the crystal (Takachiho et al. 2014).

For frequency-doubling of pulses with an average power of 1 kW, the temperature of the SHG-crystal should be stabilized. Another option for SHG is utilizing crystals with noncritical phase-matching (NCPM) which is characterized by the absence of walk-off, maximized effective nonlinearity and very wide acceptance angle. The

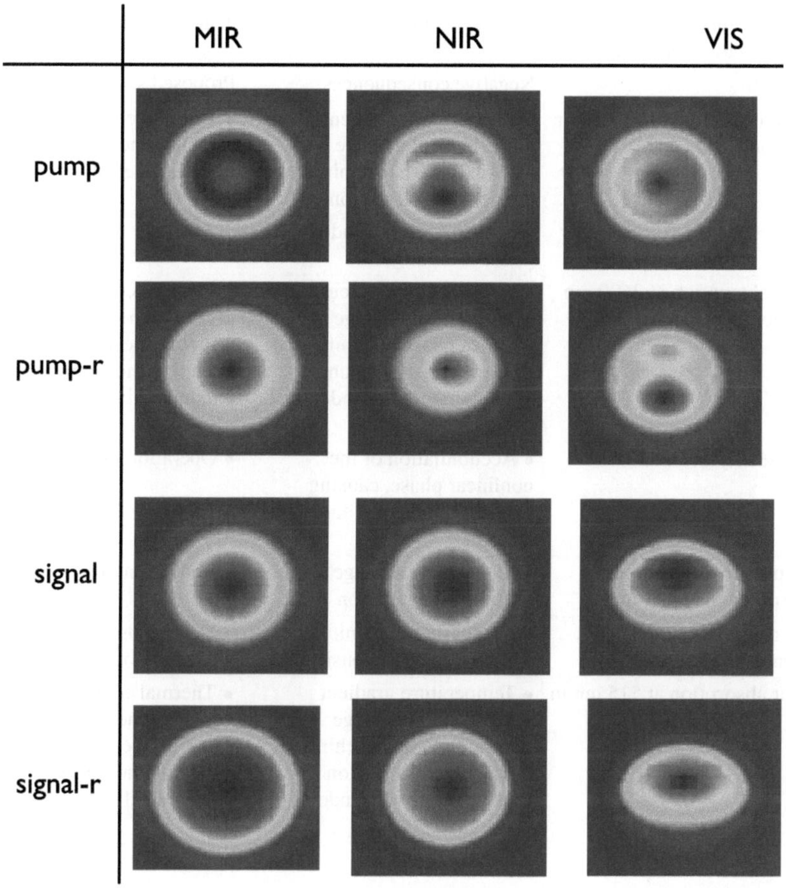

Fig. 6.6 Beam profile of the signal and pump pulses at the last (*signal-r* and *pump-r*) and the penultimate (*signal*, *pump*) OPCPA stages of three channels of the synthesizer. The amplified signal in the *MIR* and *NIR* channels preserve the Gaussian shape. The asymmetric beam profile of the signal in the *VIS* channel comes as a consequence of the large noncollinear angle between *signal* and *pump*. This effect can be avoided via pulse-front-tilt compensation

temperature bandwidth and the acceptance angle of LBO for NCPM is 4 °C and 52 mrad/cm$^{1/2}$, respectively. As for the current system, critical phase-matching is temperature and angle dependent, and using NCPM could be preferable. Table 6.2 summarizes some of the challenges with potential solutions.

Table 6.2 Summary of possible challenges for the described three-channel synthesizer and different approaches to resolve them

Challenge	Negative consequences	Proposed solution
Linear absorption at 515 nm in BBO and LBO	• Temperature gradient • Mechanical damage • Varied phase matching over the beam cross section • Temperature dependent phase matching	• Thermal control of the crystal by heating it up, utilizing special holders
Linear absorption at 2600 nm in $LiNbO_3$	• Temperature gradient • Mechanical damage • Varies phase matching over the beam cross section • Temperature dependent phase-matching	• Thermal control of the crystal by heating it up, utilizing special holders • Cutting the amplification bandwidth in OPCPA
Nonlinearity in air	• Accumulation of the nonlinear phase, causing undesired temporal and spatial changes	• Operation in vacuum
Two-photon absorption in BBO pumped at 343 nm	• Mechanical damage • Crystal degradation	• Reduction of pump intensity
BBO degradation in the presence of UV	• Decreasing transmittance with increasing intensity	• Heating up the crystal • Ejection of O_2
Linear absorption at 515 nm in LBO used for SHG	• Temperature gradient • Mechanical damage • Varied phase-matching over the beam cross section • Temperature dependent phase matching	• Thermal control of the crystal by heating it up, utilizing special holders • Using noncritical phase-matching

6.3 Field Synthesizer

Waveform-controlled light transients with a bandwidth approaching two octaves have been demonstrated at μJ energy and GW peak power levels (Hassan et al. 2012; Wirth et al. 2011; Harth et al. 2012; Manzoni et al. 2012). They allow the control of electric fields of femtosecond pulses on a sub-cycle scale (Luu et al. 2013). The described prototypical three-color few-cycle OPCPA system offers a conceptually simple route to scaling multi-octave optical waveform synthesis to the multi-terawatt regime. To this end, the three channels delivering few-cycle pulses in the VIS, NIR, and MIR spectral ranges are recombined into one beam using a set of dichroic chirped mirrors (Huang et al. 2012; Wirth et al. 2011). Due to the difficulties of dispersion management, the high demands on the chirped-mirror compressor, and the required coating for the OPCPA crystals in serial pulse synthesis (Harth et al. 2012; Herrmann et al. 2010b), a parallel-synthesis approach is chosen (Kärtner et al. 2013).

In coherent synthesis of different pulses, fine control over the spectral phase, relative phase of each pulse, and their relative timing jitter are crucial. Fluctuations and drifts in the relative timing of the recombined pulses need to be suppressed to a tiny fraction of the half-field cycle for a stable waveform resulting from the coherent superposition (Bagayev et al. 2013). To control the spectral phase, a Dazzler acousto-optic pulse shaper (Verluise et al. 2000) will be implemented in the MIR OPCPA chain, and later the negatively-chirped, amplified pulses will be compressed to their Fourier transform limit using a spatially tailored chirped-mirror compressor. For the compression of the NIR and VIS channels, a dual spatial light modulator (DSLM) (Binhammer et al. 2005) and two sets of specially designed chirped-mirrors could be utilized. This combination allows fine control over the spectral phase of each channel to minimize the residual spectral phase in each arm separately.

A passively stabilized OPCPA seed ensures a stable carrier-envelope phase, and the relative timing stability between the three arms will be ensured by a balanced optical cross-correlator (BOC) to avoid a phase shift of the waveform at the synthesis point. By utilizing a BOC, a temporal overlap with sub-100-as precision could be achieved, as in this approach the contribution of the amplitude noise of each OPCPA output on the detected error signal is suppressed (Huang et al. 2012; Manzoni et al. 2012). Besides the precise temporal overlap of pulses, precise spatial synthesis is required for wave-front matching of the pulses. As a majority of applications will employ focusing of light transients in order to attain high intensities, the OPCPA pulses will be combined in a constant-waist-width fashion that allows the temporal pulse form to remain unchanged upon propagation (Zou and Lü 2007).

The feasibility of super-octave optical waveform synthesis was recently demonstrated in the NIR-VIS-UV spectral range by seeding a three-channel (Wirth et al. 2011) and, more recently, four-channel (Luu et al. 2013) synthesizer consisting of broadband chirped mirrors with a continuum originating from a Ti:sapphire-laser-driven HCF. Implementation with an OPCPA system is also being prepared (Kärtner et al. 2013; Rossi et al. 2014; Fang et al. 2013).

Merely the adjustment of the relative timing of the three pulses emerging from the three OPCPA channels can result in a great variety of electric field wave forms on the time scale of the optical cycle. Figure 6.7 depicts a few representative waveforms that may be synthesized by first compressing the output of the individual channels to their bandwidth limit and then adding the three pulses with various time delays and relative amplitudes. Further degrees of freedom for waveform modification can be introduced by shaping the amplitude and phase of the spectra of the individual channels.

Conclusion In this section a conceptual design of an OPCPA-based synthesizer is presented and discussed. 1.6-ps pulses from a 200-mJ Yb:YAG thin-disk regenerative amplifier operating at 5 kHz are sent to a simple frequency converter module for generation of second and third harmonics. A multi-octave seed spectrum is divided to three parts, centered at 550, 1, and 2 μm. Consequently, these spectra are amplified to the total energy of 49 mJ in three OPCPA channels pumped by fundamental, second- and third-harmonic pulses of the regenerative amplifier. Each OPCPA channel supports few-cycle pulses.

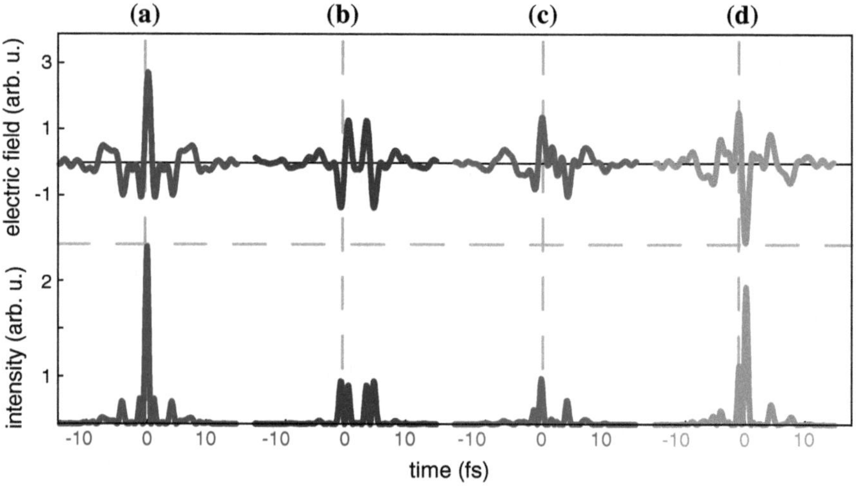

Fig. 6.7 Calculated synthesized waveforms from the three-channel OPCPA system. **a** Fourier-transform limit waveform of the three OPCPA channels. **b** Synthesized waveform generated by changing the amplitude ratio between the three channels to (1:-4/3:1), (**c**) and by changing the amplitude ratio of three different spectra to (1:0.5:1) and the delay between them to (0:-0.6 fs:1 fs). **d** Optimized waveform for HHG in terms of the highest cutoff energy of the synthesizer

Coherent combination of the three channels after their temporal compression results in a sub-cycle waveform. By adjusting the relative timing and amplitude between three arms of the synthesizer, variety of confined, sub-cycle electric fields with different shapes are achievable.

6.4 Prospects for Experiments

Multi-terawatt, multi-octave light transients are promising tools to extend the photon energy frontier of attosecond pulses to kilo-electronvolts and the boundary of the hard x-rays regime (Moulet et al. 2014). To explore the potential of keV HHG, the cutoff energy of different waveforms is evaluated by calculating the dipole response of one atom using the strong-field approximation (Lewenstein et al. 1994; Antoine et al. 1996) (see: Appendix B). Figure 6.8 compares the harmonic spectra of a 4.6 fs pulse centered at 800 nm and a 2.6 fs single-cycle pulse at 800 nm. The dipole response is calculated for a pump intensity of 50×10^{14} W/cm^2 interacting with He atoms.

Without controlling the waveform, the cutoff tends to reduce for single-cycle pulses, as the HHG process requires two half-cycles to push electrons away from their parent core and accelerate them home. Diminishing either of the half-cycles decreases the cutoff. This problem is unavoidable for a single-cycle Gaussian pulse, as the envelope of the pulse becomes shorter compared to the cycle. Controlled-transient waveforms hold promise to overcome this shortcoming, as the half-cycles of the waveform can be fully optimized by adjusting different parameters like spectral

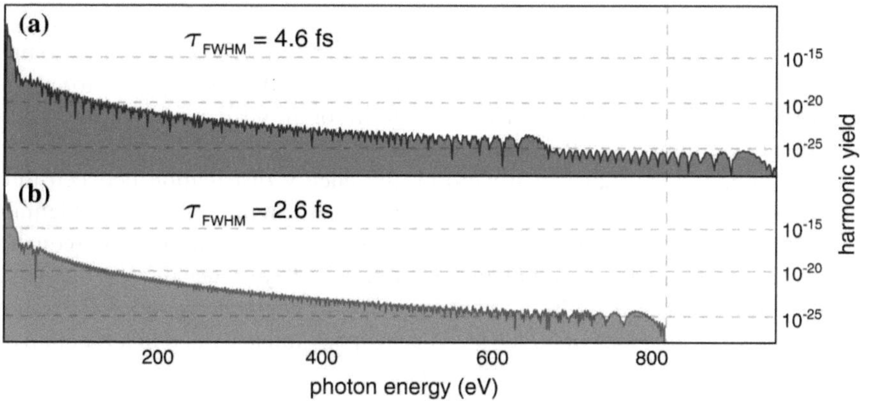

Fig. 6.8 **a** Simulated HHG spectrum of a 4.6 fs and (**b**) 2.6 fs pulse centered at 800 nm at the pump intensity of 50×10^{14} W/cm^2 in He. The cutoff decreases for the single-cycle pulse as the envelope of the pulse becomes shorter compared to the cycle

phase and amplitude of each channel, the relative temporal delay between different arms of the synthesizer, carrier envelope phase and the spectral chirp of the final waveform.

Figure 6.9 shows the influence of the carrier envelope phase on the HHG spectrum of the waveform shown in Fig. 6.7a at a peak intensity of 50×10^{14} W/cm^2 in He.

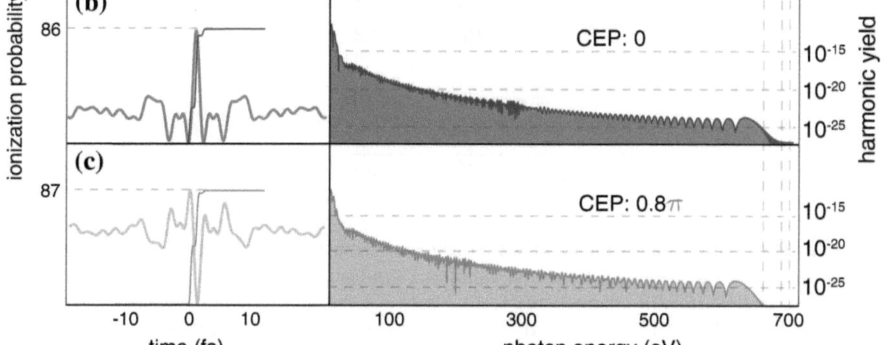

Fig. 6.9 The synthesized waveform with the CEP offset of 0.4π shows the highest HHG cutoff (**a**), compared to a waveform with carrier envelope phase offset of 0 (**b**) and 0.8π (**c**)

The highest cutoff energy is achieved for a global phase of 0.4π, as in this case the waveform contains two strong half-cycles. The first half-cycle strongly ionizes the atoms and the second one provides the force for recollision of the electrons with their parent ions. The optimized waveform in terms of the CEP offset, has the highest ionization rate compared to the other waveforms.

In the multi-octave transient waveform, the energy distribution between three channels and therefore, the shape of the spectrum plays an important role in defining the cutoff energy. Low-frequency photons extend the cutoff and higher frequency photons are helpful in the ionization process although they reduce the cutoff to lower photon energies. In Fig. 6.10 the cutoff energies for different energy ratios of the three channels of the synthesizer at $50 \times 10^{14}\,\mathrm{W/cm^2}$ peak intensity and in a He atom are

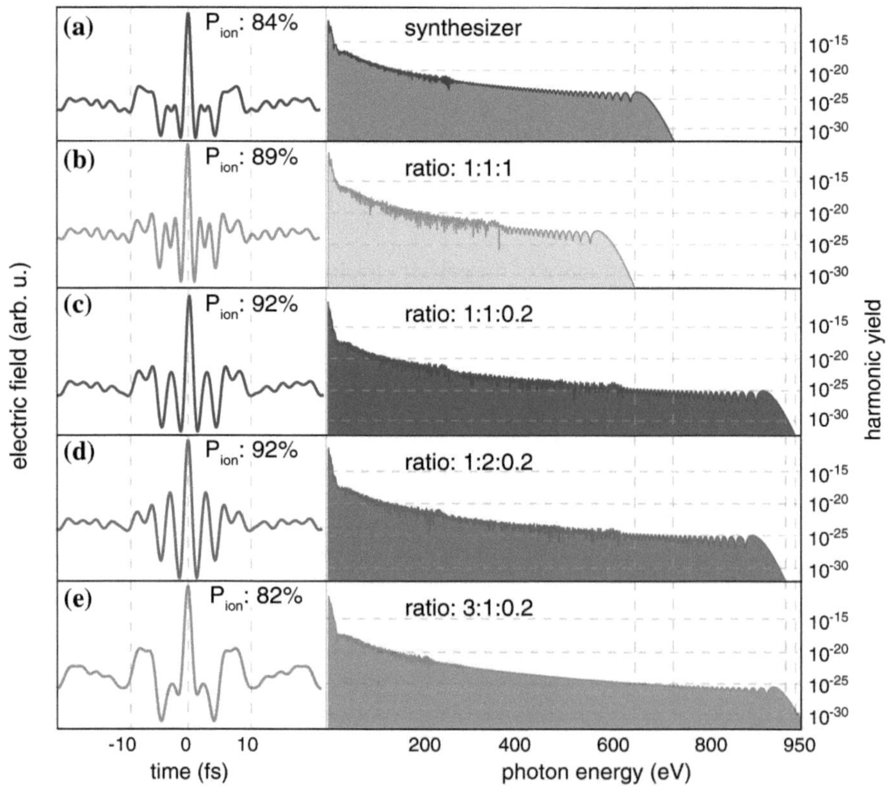

Fig. 6.10 *Right panels* the simulated HHG spectra for $50 \times 10^{14}\,\mathrm{W/cm^2}$ of peak intensity in He gas, for the energy distribution of (**a**) the simulated synthesizer in the previous section (**b**) 1:1:1 between the MIR:NIR:VIS channels (**c**) 1:1:0.2 between the MIR:NIR:VIS channels (**d**) 1:2:0.2 between the MIR:NIR:VIS channels (**d**) 1:2:0.2 between the MIR:NIR:VIS channels (**e**) 3:1:0.2 between the MIR:NIR:VIS channels. *Left panels* the corresponding transient waveform responsible for the shown spectra and their corresponding ionization probabilities. The simulation is performed for a peak intensity of $50 \times 10^{14}\,\mathrm{W/cm^2}$ and in a He atom. P_{ion}: ionisation probability

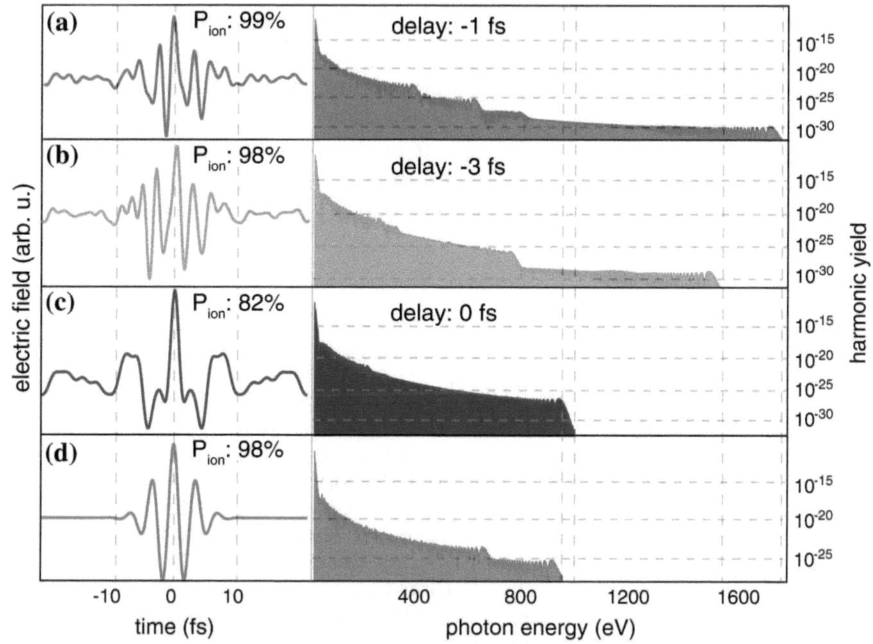

Fig. 6.11 *Right panels* the HHG spectra generated with a transient waveform of the synthesizer with energy distribution of 3:1:0.2 for VIS:NIR:MIR channels, while the VIS channel is delayed by (**a**) −1 fs (**b**) −3 fs and (**c**) 0 fs relative to the other channels. (**d**) The HHG spectrum of a 4.6 fs Gaussian pulse shown for comparison. *Left panels* the corresponding synthesized waveform that produced the HHG spectra. P_{ion}: ionisation rate

compared. Reducing the energy of the VIS channel extends the cutoff dramatically while increasing the energy of the MIR channel improves the cutoff energy. For the case of an energy ratio of 3:1:0 between the MIR:NIR:VIS channel, the cutoff decreases, emphasizing the role of the VIS channel in the ionization process.

This suggests that the cutoff can be optimized further by changing the relative delay of the VIS channel compare to the other channels in order to shift the high frequency photons in the VIS range to the leading edge of the waveform where the first half-cycle ionizes the atom. As shown in Fig. 6.11, this selectively introduced chirp leads to a substantial increase in the generated photon energies. Comparing the ionization probabilities in Figs. 6.10a, 6.11d suggests that synthesized waveforms are also superior to Gaussian pulses in terms of reaching the same cutoff energy at a lower ionization rate, as one of the fundamental limits in phase-matching high-energy photons is determined by the plasma density. We have neglected the magnetic drift as the electron in this parameter regime is not yet relativistic (Palaniyappan et al. 2006; Reiss 2014).

Conclusion Sub-1-cycle waveforms hold promise to improve the cut-off energy in the HHG process. Due to their confined sub-cycle electric field, they provide the same cut off energy in lower ionization probability compared to Gaussian pulses. This improves the phase-matching over propagation.

It is analytically shown that the cut off of the waveform driven HHG can be extended by pushing the high energy photons to the leading edge of the waveform and leaving the low-energy photons at the trailing edge of the waveform.

References

Fattahi H, Barros HG, Gorjan M, Nubbemeyer T, Alsaif B, Teisset CY, Schultze M, Prinz S, Haefner M, Ueffing M, Alismail, A,Vámos L, Schwarz A, Pronin O, Brons J, Geng XT, Arisholm G, Ciappina M, Yakovlev V.S, Kim D-E, Azzeer A M, Karpowicz N, Sutter D, Major Z, Metzger T, Krausz F (2014a) Third-generation femtosecond technology. Optica 1(1):45–63

Deng Y, Schwarz A, Fattahi H, Ueffing M, Gu X, Ossiander M, Metzger T, Pervak V, Ishizuki H, Taira T, Kobayashi T, Marcus G, Krausz F, Kienberger R, Karpowicz N (2012) Carrier-envelope-phase-stable, 1.2 mJ, 1.5 cycle laser pulses at 2.1 μm. Opt Lett 37(23):4973–4975

Hong KH, Huang SW, Moses J, Fu X, Lai CJ, Cirmi G, Sell A, Granados E, Keathley P, Franz X (2011) High-energy, phase-stable, ultrabroadband kHzOPCPA at 2 .1 um pumped by a picosecond cryogenic Yb : YAG laser. Opt Exp 19(16):15538–15548

Kartashov D, Ališauskas S, Pugžlys A, Voronin A, Zheltikov A, Petrarca M, Béjot P, Kasparian J, Wolf J-P, Baltuška A (2012) White light generation over three octaves by femtosecond filament at 3.9 μm in argon. Opt Lett 37(16):3456–3458

Silva F, Austin DR, Thai A, Baudisch M, Hemmer M, Faccio D, Couairon A, Biegert J (2012) Multi-octave supercontinuum generation from mid-infrared filamentation in a bulk crystal. In: Nat commun 3:8071–8075

Schwarz A (2014) Few-cycle phase-stable infrared OPCPA. (Ph.D thesis)

Voronin AA, Mikhailova JM, Gorjan M, Major Z, Zheltikov AM (2013) Pulse compression to subcycle field waveforms with split-dispersion cascaded hollow fibers 38(21):4354–4357

Cotel A, Jullien A, Forget N, Albert O, Chériaux G, Le Blanc C (2006) Nonlinear temporal pulse cleaning of a 1-μm optical parametric chirped-pulse amplification system. Appl Phys B 83(1):7–10

Homann, C., Bradler, M., Förster, M., Hommelhoff, P., and Riedle, E. (2012). Carrier-envelope phase stable sub-two-cycle pulses tunable around 1.8 μm at 100 kHz. Opt Lett 37(10):1673–1675

Seidel M, Brons J, Fedulova E, Pervak V, Apolonski A, Pronin O, Krausz F (2014) High-Power Few-Cycle Pulse Generation by Spectral Broadening in Bulk Material. Postdeadline Paper Digest, Washington, OSA, STH5C.9

Tate J, Auguste T, Muller H, Salières P, Agostini P, DiMauro L (2007) Scaling of wave-packet dynamics in an intense midinfrared field. Phys Rev Lett 98(1):0139011–0139014

Banks PS, Feit MD, Perry MD (2002) High-intensity third-harmonic generation. J Opt Soc Am B 19(1):102–118

Herrmann D, Homann C, Tautz R, Scharrer M, Russell PSJ, Krausz F, Veisz L, Riedle E (2010a) Approaching the full octave: noncollinear optical parametric chirped pulse amplification with two-color pumping. Opt Exp 18(18):18752–18762

Rothhardt J, Demmler S, Hädrich S, Peschel T, Limpert J, Tünnermann A (2013) Thermal effects in high average power optical parametric amplifiers. Opt Lett 38(5):763–765

Prandolini MJ, Riedel R, Schulz M, Hage A, Tavella F, Jena D (2014) Design considerations for a high power, ultrabroadband optical parametric chirped-pulse amplifier. Opt Exp 22(2):1594–1607

Takachiho K, Yoshimura M, Takahashi Y, Imade M, Sasaki T, Mori Y (2014) Ultraviolet laser-induced degradation of CsLiB6O10 and β-BaB2O4. Opt Mater Exp 4(3):559–567
Hassan MT, Wirth A, Grguraš I, Moulet A, Luu TT, Gagnon J, Pervak V, Goulielmakis E (2012) Invited article: attosecond photonics: synthesis and control of light transients. Rev Sci Instrum 83(11):111301–111319
Wirth A, Hassan MT, Grguras I, Gagnon J, Moulet A, Luu TT, Pabst S, Santra R, Alahmed ZA, Azzeer AM, Yakovlev VS, Pervak V, Krausz F, Goulielmakis E (2011) Synthesized light transients. Science (New York) 334(6053):195–200
Harth A, Schultze M, Lang T, Binhammer T, Rausch S, Morgner U (2012) Two-color pumped OPCPA system emitting spectra spanning 1.5 octaves from VIS to NIR. Opt Exp 20(3):3076–3081
Manzoni C, Huang S-W, Cirmi G, Farinello P, Moses J, Kärtner FX, Cerullo G (2012) Coherent synthesis of ultra-broadband optical parametric amplifiers. Opt Lett 37(11):1880–1882
Luu TT, Hassan MT, Moulet A, Razskazovskaya O, Kaprowicz N, Pervak V, Krausz F, Goulielmakis E (2013) Isolated optical attosecond pulses–OSA Technical Digest (online) (CLEO: 2013). Opt Soc Am (California) QF1C.6
Huang S-W, Cirmi G, Moses J, Hong K-H, Bhardwaj S, Birge JR, Chen L-J, Kabakova IV, Li E, Eggleton BJ, Cerullo G, Kärtner FX (2012) Optical waveform synthesizer and its application to high-harmonic generation. J Phys B: At Mol Opt Phys 45(7):074009–074114
Herrmann D, Homann C, Tautz R, Scharrer M, Russell PSJ, Krausz F, Veisz L, Riedle E (2010b) Approaching the full octave: noncollinear optical parametric chirped pulse amplification with two-color pumping. Opt Exp 18(18):18752–18762
Kärtner FX, Mücke O, Cirmi G, Fang S, Chia S-H, Manzoni C, Farinello P, Cerullo G (2013) High energy sub-cycle optical waveform synthesizer. Advanced Solid-State Lasers Congress. Ed. by G. Huber and P. Moulton.Washington, D.C., OSA, AW2A.1
Bagayev SN, Trunov VI, Pestryakov EV, Leschenko VE, Frolov SA, Vasiliev VA (2013) High-intensity femtosecond laser systems based on coherent combining of optical fields. Opt Spectrosc 115(3):311–319
Verluise F, Laude V, Cheng Z, Spielmann C, Tournois P (2000) Amplitude and phase control of ultrashort pulses by use of an acousto-optic programmable dispersive filter: pulse compression and shaping. Opt Lett 25(8):575–577
Binhammer, T., Rittweger, E., Ell, R., Kartner, F., and Morgner, U. (2005). Prism-based pulse shaper for octave spanning spectra. IEEE J Quant Electron 41(12):1552–1557
Zou Q, Lü B (2007) Propagation properties of ultrashort pulsed beams with constant waist width in free space. Opt Laser Technol 39(3):619–625
Rossi GM, Cirmi G, Fang S, Chia S-H, Muecke OD, Kärtner F, Manzoni C, Farinello P, Cerullo G (2014) Spectro-temporal characterization of all channels in a sub-optical-cycle parametric waveform synthesizer (CLEO: 2014). Washington, D.C., OSA, SF1E.3
Fang S, Cirmi G, Chia S-H, Mücke OD, Kärtner FX, Manzoni C, Farinello P, Cerullo G (2013) Multi-mJ parametric synthesizer generating two-octave-wide optical waveforms. In: Conference on Lasers and Electro-Optics Pacific Rim. Kyoto: Optical Society of America, $WB3_1$
Moulet A, Tosa V, Goulielmakis E (2014) Coherent kiloelectronvolt x-rays generated by subcycle optical drivers: a feasibility study. Opt Lett 39(21):6189–92
Lewenstein M, Balcou P, Ivanov M, L'Huillier A, Corkum P (1994) Theory of high-harmonic generation by low-frequency laser fields. Phys Rev A 49(3):2117–2132
Antoine P, L'Huillier A, Lewenstein M, Saliéres P, Carré B (1996) Theory of high-order harmonic generation by an elliptically polarized laser field. Phys Rev A 53(3):1725–1745
Palaniyappan S, Ghebregziabher I, DiChiara A, MacDonald J, Walker B (2006) Emergence from nonrelativistic strong-field rescattering to ultrastrongfield laser-atom physics: a semiclassical analysis. Phys Rev A 74(3):033403–11
Reiss HR (2014) The tunneling model of laser-induced ionization and its failure at low frequencies, pp 1–19. arXiv:1403.0568v1

Erny C, Moutzouris K, Biegert J, Kühlke D, Adler F, Leitenstorfer A, Keller U (2007) Mid-infrared difference-frequency generation of ultrashort pulses tunable between 3.2 and 4.8 mm from a compact fiber source. Opt Lett 32(9):1138–1140
Chalus O, Bates PK, Smolarski M, Biegert J (2009) Mid-IR shortpulse OPCPA with micro-Joule energy at 100kHz. Opt Express 17(5):3587–3594

Chapter 7
Conclusion and Outlook

The first-generation of femtosecond technology relied on dye lasers with MW-scale peak power and average powers up to 100 mW. The spectral bandwidth of these sources was confined to a narrow range between 600 and 900 nm.

The second-generation femtosecond lasers, driven by chirped pulse amplification (CPA) in broadband solid-state laser media, boosted the peak power of lasers to the multi-terawatt level. At this peak power lasers provide enough intensity to efficiently harvest $\chi^{(2)}$ and $\chi^{(3)}$ nonlinearities and reach extreme nonlinearity at low repetition rates but are still limited in terms of reaching high peak power at high repetition rates.

Based on short-pulse-pumped optical parametric chirped pulse amplification (OPCPA), third-generation femtosecond technology allows boosting the peak and average power of coherent femtosecond light sources to the multi-terawatt level with hundreds of Watts average power, at a MHz repetition rate.

The OPCPA systems designed and realized in this thesis are pumped with near-1-ps Yb:YAG thin-disk lasers and have a modular and robust architecture offering different modes of operation. From a high-power, widely tunable source to few-cycle sources at different carrier wavelengths spanning from visible to mid-infrared or super-octave, subcycle light transients.

Crucial for the performance of this concept is the availability of broadband, high-contrast, CEP-stable seed pulses and powerful and reliable pump sources:

In Chap. 3 of this work, different techniques for the generation of coherent, broadband spectra suitable for seeding OPCPA systems were theoretically and experimentally investigated. Comparison of several techniques revealed that the smooth spectrum and good temporal contrast made cross-polarized wave generation (XPW) and difference frequency generation (DFG) the most favorable seed generation schemes. Additionally, the DFG concept offers intrinsic phase-lock between the carrier wave and the intensity envelope (CEP stability). However, the efficiency of these processes is low, for XPW due to the weak third-order nonlinearity and in broadband-DFG because of the limit phase-matching requirements impose on the nonlinear crystal

H. Fattahi, *Third-Generation Femtosecond Technology*,
Springer Theses, DOI 10.1007/978-3-319-20025-5_7

thickness. Here, it is experimentally shown that the efficiency can be doubled by implementing a cascaded XPW setup. Optimization of the DFG process up to 10 % efficiency is achieved by careful projection of the input signal on both the ordinary and extraordinary axis of the nonlinear crystal, making DFG the method of choice for OPCPA seed generation.

Chapter 4 investigates Yb:YAG thin-disk regenerative amplifiers as the most promising pump sources for high-energy, high-power OPCPA systems. To fully exploit the capabilities of these sources, the underlying dynamics of chirped pulse amplification including the chaotic amplification scheme in an ytterbium-based thin-disk amplifier is examined. It was demonstrated that with adequate input seed energy and optimized cavity design, the nonlinear amplification dynamics and the chaotic regime in regenerative amplifiers are in balance and a safe operation mode is achieved. Yielding <1 % amplitude stability renders these systems ideal pump sources for nonlinear amplification schemes like OPCPA.

With broadband seed pulses and the reliable, 20 mJ Yb:YAG thin-disk regenerative amplifier in place, different techniques for further optimization of the conversion efficiency and spectral bandwidth engineering in an OPCPA system were experimentally and theoretically studied (Chap. 5). It was shown that by reusing the residual pump energy in a 3-stage OPCPA system and employing BBO crystals as the nonlinear medium, a pump-to-signal conversion efficiency of 30 % is feasible while the amplified spectrum supports sub-6-fs pulses.

In a conventional OPCPA system, a balance between amplification bandwidth and conversion efficiency has to be found. Necessarily, the seed pulse duration has to be shorter than the pump pulses. Consequently only a fraction of the pump pulse energy can be used for amplification, limiting the overall efficiency. This can be resolved by controlling the deposition of pump energy along the temporal profile of the heavily chirped seed pulse. Chapter 5 presents the theoretical and experimental evidence of this new scheme, promising unprecedented efficiency.

The discussion is concluded by a theoretical outline and experimental verification of a technique to engineer the OPCPA amplification gain bandwidth. Based on this, the bandwidth of a 3-stage OPCPA system was extended beyond the phase-matching limitations by utilizing different gain media (LBO and BBO in this case). Amplification of near-1-cycle pulses, at 3 kHz repetition rate and centered at 1 μm to 1.8 mJ was demonstrated, parameters that have been unachievable with 2nd generation femtosecond technology.

All of the above presented findings are utilized in Chap. 6, for the full design study of the multi-TW, multi-octave, 200-W-scale, OPCPA-based field synthesizer currently under development at the Laboratory of Extreme Photonics (LEX) in Garching.

The fundamental, second and third harmonics of a 200 mJ, 1 kW Yb:YAG thin-disk laser will be used to pump three OPCPA channels covering the visible to mid-infrared wavelength range. The superposition of the few-cycle pulses from the individual channels will allow synthesis of sub-cycle waveforms. Analytical methods were employed to identify the ideal waveform that maximises the cutoff energy of the generated photons in the high harmonic generation (HHG) process. With the optimized waveform seeding HHG, the laser system is predicted to become a new source of attosecond keV X-ray pulses.

Such a laser will equip scientists with a unique tool to explore and scrutinize a new era of physics so far deemed beyond reach. The attempts to better understand and control the underlying principles of lasers in the single cycle limit, discussed in this thesis, are small contributions to the next chapter of the fascinating story of photons.

Appendix A
In Situ Spectral Interferometry as a Pulse Compression Tool

Compressing the output of a laser or an OPA requires that both the pulse and the compressor are free of complicated phase features, unless they are equal and opposite. However, once a complicated structure is detected in the final pulse, the question still remains if it is coming from the laser or the compressor. For example, a split pulse could arise either through a strongly amplified reflection, or through an unwanted ripple in the group delay curve of a set of chirped mirrors. Finding the culprit requires a way of measuring at least one of them independently.

FROG and other pulse measurement techniques are based on nonlinear effects, and cannot be easily used for an uncompressed pulse. A simple and effective way to characterize the chirped-mirror compressor is measuring the total group delay response of the compressor using interferometry, which is a linear effect. This method does not require the pulse to be compressed to perform a measurement, can be done in a single shot and requires only a spectrometer capable of detecting the spectrum of the laser or OPA itself.

In order to characterize the chirped-mirror compressor used in Sect. 5.3, the seed pulse of the OPCPA system was sent to the chirped-mirror compressor. A 4-mm-thick CaF_2 plate beam splitter was placed before the compressor, reflecting a reference pulse which bypassed the compressor. After the rest of the light was transmitted through the compressor, it was recombined with the reference light (which passed through a variable time delay), using an identical glass plate beam splitter (Fig. A.1). The combined beams were sent to a spectrometer so that their spectral interference pattern could be observed.

These patterns, accompanied by reference spectra from the two arms independently, are sufficient to completely reconstruct the group delay curve of the compressor.

H. Fattahi, *Third-Generation Femtosecond Technology*,
Springer Theses, DOI 10.1007/978-3-319-20025-5

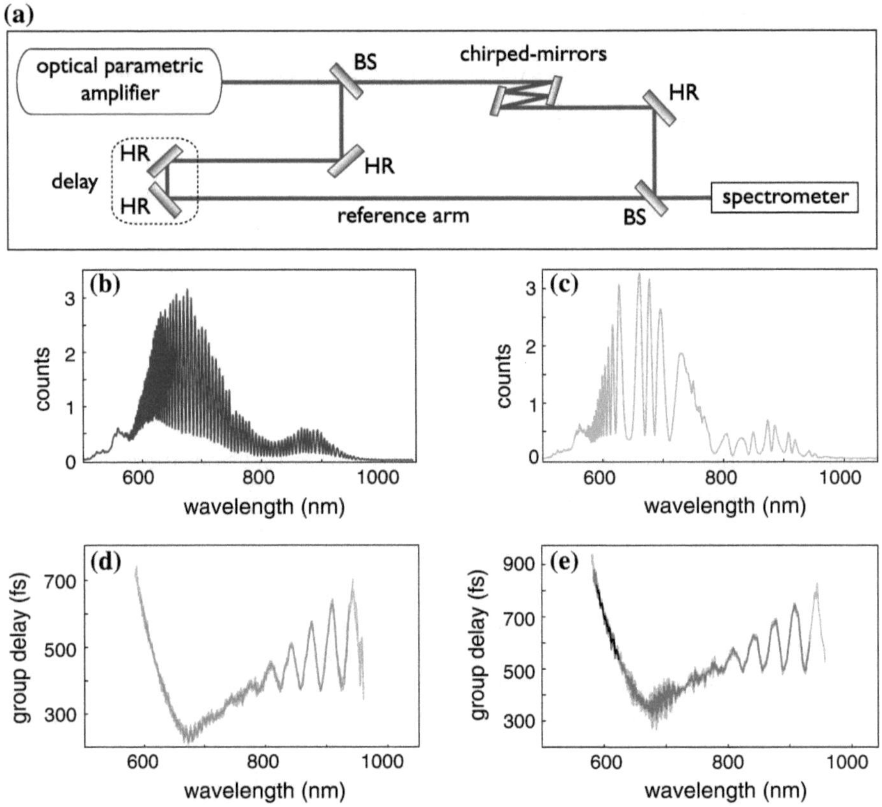

Fig. A.1 **a** Layout of the spectral interferometer setup; *BS* beam splitter; *HR* high reflector. **b** Recorded spectrum with reference significantly temporally separated from signal. **c** Spectrum with a smaller path length difference. **d** Reconstructed group delay curve from (**b**). **e** Reconstructed group delay curve from (**c**). 12 mm of glass was inserted in the compressor to remove some of the negative GDD so that the high-order terms were more visible

The spectral intensity of the two pulses after interference is described as:

$$\begin{aligned} I(\omega) &\propto \left|\tilde{E}_A(\omega) + \tilde{E}_B(\omega)\right|^2 \\ &\propto \left|\tilde{E}_A(\omega)\right|^2 + \left|\tilde{E}_B(\omega)\right|^2 + 2\left|\tilde{E}_A(\omega)\right|\left|\tilde{E}_B(\omega)\right| \cos\phi_{rel}(\omega) \\ &= I_A(\omega) + I_B(\omega) + 2\sqrt{I_A I_B}\cos\phi_{rel}(\omega). \end{aligned}$$

The cosine of the spectral phase can be calculated from the measurement:

$$\cos\phi_{rel}(\omega) = \frac{I(\omega) - I_A(\omega) - I_B(\omega)}{2\sqrt{I_A I_B}},$$

and by using the Hilbert transform:

$$\sin \phi_{rel}(\omega) = \frac{1}{\pi} \text{p.v.} \int_{-\infty}^{\infty} \frac{\cos \phi_{rel}(\omega)}{\omega - \omega'} d\omega'.$$

Therefore the spectral phase is:

$$\phi_{rel} = \text{atan2}(\sin \phi_{rel}(\omega), \cos \phi_{rel}(\omega)).$$

Measured interference patterns and their corresponding group delay curves for two different time delays are shown in Fig. A.1.

Conclusion Spectral interference is an extremely useful technique for characterizing a compressor without having to perform a pulse retrieval. Because it relies on linear optics, it can be performed with any beam light source without having to approximately compress the pulse beforehand. The simple setup and the single shot measurement makes it possible to optimize the reflection angles of a set of mirrors.

Appendix B
Toolbox and Ideas

OPCPA simulations The numerical simulation of the tunable OPCPA system, the multi-channel synthesizer and case studies presented in Chap. 2 were performed using the SISYFOS code, developed by Dr. Gunnar Arisholm.

SISYFOS uses a Fourier-space method to simulate second-order nonlinear interactions. In this method, each beam is decomposed into plane-wave eigenmodes and the coupled differential equations are solved for the slowly varying eigenmode amplitudes (Arisholm 1997). The code is capable of handling wide bandwidth, linear and two-photon absorption, thermal effects, nonlinear refraction, parasitic processes and noncollinear beams with possibly tilted pulse fronts. In the numerical studies in this thesis, thermal effects are not included. and the Sellmeier coefficients are taken from (Kato 1994; Zelmon et al. 1997; Zhang et al. 2000).

Nick's photonics library The numerical studies of difference frequency generation, cross-polarized wave generation, spectral interferometry, and chaotic behavior in Yb-doped lasers were done using Nick's photonics library. This marvelous library was developed by genius Dr. Nicholas Karpowicz and in the case of cross-polarized wave generation calculation, it was modified in the least clever way by the author of this thesis!

HHG simulation For numerical studies of the single-dipole HHG response in Chap. 6, a code based on the Lewenstein model (Lewenstein et al. 1994) and developed by Dr. Valer Tosa was used (Moulet et al. 2014).

TAPIR idea The idea explained in Sect. 5.2, was conceived by Dr. Yunpei Deng (Deng and Krausz 2013).

Patent Number: EP2742384, US14/0293405.

Title: "Method and device for optical parametric chirped pulse amplification."

Inventors: Yunpei Deng, Ferenc Krausz.

H. Fattahi, *Third-Generation Femtosecond Technology*,
Springer Theses, DOI 10.1007/978-3-319-20025-5

References

Arisholm G (1997) General numerical methods for simulating secondorder nonlinear interactions in birefringent media. J Opt Soc Am B 14(10):2543–2549

Deng Y, Krausz F (2013) Method and device for optical parametric chirped pulse amplification

Kato K (1986) Second-harmonic generation to 2048 Åin ß-Ba2O4. IEEE J Quant Electron 22(7):1013–1014

Lewenstein M, Balcou P, Ivanov M, L'Huillier A, Corkum P (1994) Theory of high-harmonic generation by low-frequency laser fields. Phys Rev A 49(3):2117–2132

Moulet A, Tosa V, Goulielmakis E (2014) Coherent kiloelectronvolt x-rays generated by subcycle optical drivers: a feasibility study. Opt Lett 39(21):6189–6192

Zelmon DE, Small DL, Jundt D (1997) Infrared corrected Sellmeier coefficients for congruently grown lithium niobate and 5 mol.% magnesium oxide—doped lithium niobate. J Opt Soc Am B 14(12):3319–3322

Zhang D, Kong Y, Zhang J (2000) Optical parametric properties of 532-nm-pumped beta-barium-borate near the infrared absorption edge. Opt Commun 184 (5–6):485–491

Appendix C
Simulation Details of Fig. 1.2

Table C.1 summarizes the parameters of the simulations performed for OPCPA systems depicted with red triangles in Fig. 1.2. In all calculations, a pump-to-seed energy ratio of 10^3 is considered. The pump pulses have a Gaussian profile in time and space. For the temporal profile of the signal, a Gaussian distribution of the fourth order is assumed. LBO crystal with a noncollinear pump to signal angle of 1.1° and a phase-matching angle of 14.4° is used as a nonlinear medium. The d_{eff} value is 0.819×10^{-12} pm/V. In all simulations, three-stages of OPCPA is considered while in the last stage the residual pump energy of the second stage is recycled. The amplified spectrum in all cases is centered at 1 μm.

Table C.1 Summary of the simulated OPCPA systems depicted in Fig. 1.2

300 W, ps thin-disk regenerative amplifier at 10 kHz	
Pump laser	30 mJ, 10 kHz, 1.6 ps
Second harmonic	21 mJ, 70 % efficiency
Pump peak intensity	110 GW/cm^2
Amplified energy/pulse duration	6.1 mJ/5 fs
Compact diode-pumped 1.1 kW Yb:YAG Innoslab femtosecond amplifier	
Pump laser	1.38 mJ, 800 kHz, 7.3 ps
Second harmonic	700 μJ, 50 % efficiency
Pump peak intensity	40.7 GW/cm^2
Amplified energy/pulse duration	128 μJ/5.24 fs
200 mJ, ps X-CPA thin-disk regenrative amplifier at 5 kHz	
Pump laser	200 mJ, 5 kHz, 1.6 ps
Second harmonic	140 mJ, 70 % efficiency
Pump peak intensity	110 GW/cm^2
Amplified energy/pulse duration	40 mJ/5 fs
Envisioned 2 J, thin-disk booster at 10 kHz	
Pump laser	2 J, 10 kHz, 1.6 ps
Second harmonic	1.4 J, 70 % efficiency
Pump peak intensity	110 GW/cm^2
Amplified energy/pulse duration	345 mJ/5 fs

H. Fattahi, *Third-Generation Femtosecond Technology*,
Springer Theses, DOI 10.1007/978-3-319-20025-5

Appendix D
1-ps-Pumped Supercontinuum Generation

In a preliminary experiment, broadband supercontinuum was generated by focusing a small fraction of a 800 fs pulses, centered at 1030 nm (Klingebiel et al. 2011) to a 20 mm-long TGG crystal.

Single filament, stable regime of operation for the input pulse energies of 70–150 μJ was achieved. In a 4mm-long KGW crystal the generated spectrum shows a different behavior which is originated from Raman effect (Fig. D.1a, b). Focusing the second harmonic of the laser in 4 mm YAG and 20 mm TGG, stable supercontinuum, with the cut-off below 500 nm was generated. As the photon energies are higher in this case, the damage of the crystal happened easier (Fig. D.1c). Simultaneous focusing of the pulses at 1030 and 515 nm resulted in a broadband spectrum covering spectral component from 470–1400 nm. These supercontinuum are not scalable in energy, as the generated filament in bulk can carry a certain amount of energy and for peak power above critical power Eq. (D.1), multi-filamentation happens.

$$P_{cr} = \frac{\pi(0.61)^2\lambda_0^2}{8n_0n_2}. \tag{D.1}$$

Their carrier envelope phase is not stable but they preserve the CEP-stability of the pump pulse. Anti-stokes and stokes wings of such supercontinuum have a well-behaved spectral phase although it is distorted at the vicinity of the pump wavelength (Cerullo et al. 1997; Siddiqui et al. 2009), makes it difficult to compress a full supercontinuum spectrum (Antipenkov et al. 2011; Riedle et al. 2000). It is shown recently that the magnitude of this phase jump scales with the pulse duration of the the pump pulses (Harth et al. 2012).

H. Fattahi, *Third-Generation Femtosecond Technology*,
Springer Theses, DOI 10.1007/978-3-319-20025-5

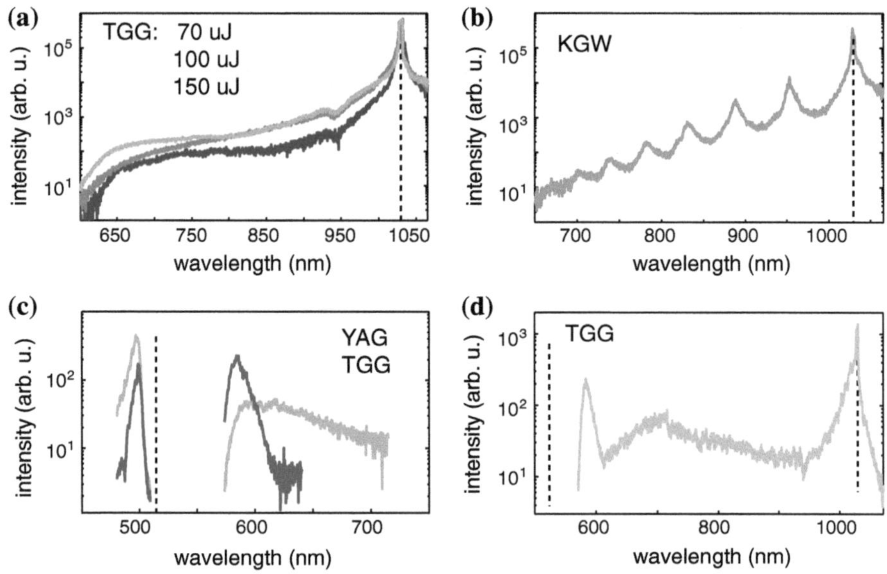

Fig. D.1 **a** Anti-stoke wing of the generated supercontinuum pumped by 1030-nm, 800-fs pulses in 20-mm TGG crystal and **b** the observed Raman effect in a 4-mm KGW crystal. **c** Supercontinuum generation in 4-mm YAG and 20-mm TGG crystal pumped by 800-fs pulses and measured after a notch filter at 515-nm. **d** ω-2ω pumped supercontinuum in 20-mm TGG. A long-pass filter at 515-nm and a short-pass filter at 1030-nm is used to filter out the pump beam

References

Antipenkov R, Varanavičius A, Zaukevičius A, Piskarskas A (2011) Femtosecond Yb:KGW MOPA driven broadband NOPA as a frontend for TW few-cycle pulse systems. Opt Exp 19(4):3519–3524

Cerullo G, Nisoli M, De Silvestri S (1997) Generation of 11 fs pulses tunable across the visible by optical parametric amplification. Appl Phys Lett 71(25):3616–3618

Harth A, Schultze M, Lang T, Binhammer T, Rausch S, Morgner U (2012) Two-color pumped OPCPA system emitting spectra spanning 1.5 octaves from VIS to NIR. Opt Exp 20(3):3076–3081

Klingebiel S, Wandt C, Skrobol C, Ahmad I, Trushin SA, Major Z, Krausz F, Karsch S (2011) High energy picosecond Yb:YAG CPA system at 10 Hz repetition rate for pumping optical parametric amplifiers. Opt Exp 19(6):5357–5363

Riedle E, Beutter M, Lochbrunner S, Piel J, Schenkl S, Spörlein S, Zinth W (2000) Generation of 10 to 50 fs pulses tunable through all of the visible and the NIR. Appl Phys B 71(3):457–465

Siddiqui AM, Cirmi G, Brida D, Kärtner FX, Cerullo G (2009) Generation of <7 fs pulses at 800 nm from a blue-pumped optical parametric amplifier at degeneracy. Opt Lett 34(22):3592–3594

MIX
Papier aus verantwortungsvollen Quellen
Paper from responsible sources
FSC® C105338

If you have any concerns about our products,
you can contact us on
ProductSafety@springernature.com

In case Publisher is established outside the EU,
the EU authorized representative is:
Springer Nature Customer Service Center GmbH
Europaplatz 3, 69115 Heidelberg, Germany

Printed by Libri Plureos GmbH
in Hamburg, Germany